Björn Gottfried

Shape from Positional-Contrast

Bildwissenschaft

Herausgegeben von
Klaus Sachs-Hombach und Klaus Rehkämper

Zunehmend werden unsere Erfahrungen und Erkenntnisse durch Bilder vermittelt und geprägt. In kaum zu überschätzender Weise halten Bilder Einzug in Alltag und Wissenschaft. Gemessen an der Bedeutung, die bildhaften Darstellungen mittlerweile zugeschrieben wird, erstaunt jedoch die bisher ausgebliebene Institutionalisierung einer allgemeinen Bildwissenschaft.

Mit dieser Buchreihe möchten die Herausgeber einen transdisziplinären Rahmen für die Bemühungen der einzelnen mit Bildern beschäftigten Fachdisziplinen zur Verfügung stellen und so einen Beitrag zum Entstehen einer allgemeinen Bildwissenschaft leisten.

Björn Gottfried

Shape from Positional-Contrast

Characterising Sketches with Qualitative Line Arrangements

Deutscher Universitäts-Verlag

Bibliografische Information Der Deutschen Nationalbibliothek
Die Deutsche Nationalbibliothek verzeichnet diese Publikation in der
Deutschen Nationalbibliografie; detaillierte bibliografische Daten sind im Internet über
<http://dnb.d-nb.de> abrufbar.

Band 18 der Reihe Bildwissenschaft, die bis 2001 im Scriptum Verlag, Magdeburg erschienen ist.

Dissertation Universität Bremen, 2005

1. Auflage Januar 2007

Lektorat: Brigitte Siegel / Frauke Schindler

Der Deutsche Universitäts-Verlag ist ein Unternehmen von Springer Science+Business Media.
www.duv.de

Umschlaggestaltung: Regine Zimmer, Dipl.-Designerin, Frankfurt/Main
Gedruckt auf säurefreiem und chlorfrei gebleichtem Papier
Printed in Germany

ISBN 978-3-8350-6070-8

Acknowledgements

The work presented in this book is the result of my dissertation. I would like to thank my advisor Prof. Dr. Christoph Schlieder, Universität Bamberg, as well as all other members of the disputation board, namely Prof. Dr. Otthein Herzog, Prof. Dr. Christian Freksa, Prof. Dr. Kerstin Schill, Sebastian Hübner, and Ole Osterhagen.

I accomplished my doctoral thesis between spring 2002 and spring 2005 during my stay at the Centre for Computing Technologies, Universität Bremen. I am grateful that the head of this institute, Prof. Dr. Otthein Herzog, gave me both the opportunity and the material support in order to conduct my research. In several discussions I received valuable feedback from many of my colleagues at the Artificial Intelligence Research Group. In particular, I would like to thank in alphabetical order Andreas Lattner, Arne Schuldt, Dr. Thomas Vögele, and Dr. Tom Wetjen.

I should also like to thank the Bamberger Naturkundemuseum for making available their historical fruit collection, and Mona Hess, Universität Bamberg, for digitising this collection.

Preface

Graphical queries for the purpose of searching for pictorial information are of growing interest in areas where pictures provide valuable information, including, for instance, design, architecture, and engineering. Sketching graphical queries is a natural way of revealing the visual appearance of objects one has in mind. The problem which arises is identifying necessary shape properties of sketches, that is, those properties which are not accidental but are necessary for specifying a particular object property. This problem arises in particular because sketches are imprecise, and often distorted by the artistic limitations of the sketcher.

From the theoretical point of view the concept of *pictorial space* applies. In this context, new concepts are required, in particular for dealing with imprecise shape information in the plane. Taking into account constraints imposed by pictorial space, a relation algebra of intersection-free relations is proposed, which allows reasoning about qualitative line arrangements in the plane. The theory is further developed for characterising polygons, by deriving a number of qualitative properties which aid in describing polygons qualitatively. Applying this theory, only line arrangements which are both readily sketched and easily perceivable are considered to be different. The notion of positional-contrast is introduced, which points out that the particular arrangements of line segments, i.e. their positions relative to each other, provide an expressive means of characterising necessary shape properties. The method developed in this work is applied in using graphical queries to search for historical objects. Specifying objects graphically, it is shown that the new method is capable of dealing with imprecise sketches by describing necessary shape properties using qualitative line arrangements, i.e. by taking into account positional-contrast.

Overall, the investigations which are carried out in this work pertain to the field of knowledge representation in artificial intelligence. The representation developed here sticks to ideas developed in the field of *qualitative spatial reasoning* — one of its principal goals being the representation of commonsense knowledge about the physical world. As is common practice in this field, knowledge about the domain in hand is made explicit in order to arrive at efficient reasoning methods. Here, we confine ourselves to a representation of shape information and make explicit knowledge about our object of research, namely about imprecise shape information in picture space.

Contents

List of Figures

List of Tables

Part I

Introduction

Chapter 1

Sketches

Pictorial information is frequently used as a means of depicting both spatial and non-spatial knowledge:

1. designers develop new ideas by making sketches;
2. architects, city planners, and landscapers outline ground plans;
3. cartographers construct maps of geographic space;
4. engineers use graphical languages for modelling concepts and processes;
5. scientists visualise their experiments and observations;
6. several occupational groups make use of image databases.

The considerable increase in information available to us, especially pictorial information, poses a problem: how do we find one particular document among a vast number? An especially complex sub-problem pertains to freehand sketches, as in the cases mentioned above. A city planner may be interested in particular configurations of geographical objects, drawing them and comparing the drawings with maps. Designers gather a great many sketches over the years, and they are sometimes interested in finding sketches which have been drawn in the past. Both city planners and designers would benefit from a system which can automatically search for documents which are similar to a query sketch. Indeed, digitised recordings in all sorts of areas are taken for granted today and have sometimes even become an indispensable source, for example, in archaeology, history, and the art trade. The search for particular objects is becoming commonplace and requires sophisticated techniques which cope with queries the user specifies graphically in order to express the visual appearance of objects which are difficult or impossible to specify verbally. Figure 1.1 shows the sketch of an object with specific properties and an image which contains an object similar to that sketch. Some of these properties may be specifiable in language but our vocabulary is both ambiguous and limited — language cannot keep pace with drawings.

Freehand sketches are different from photos or precise graphics, because they convey objects only schematically, and we have to distinguish carefully

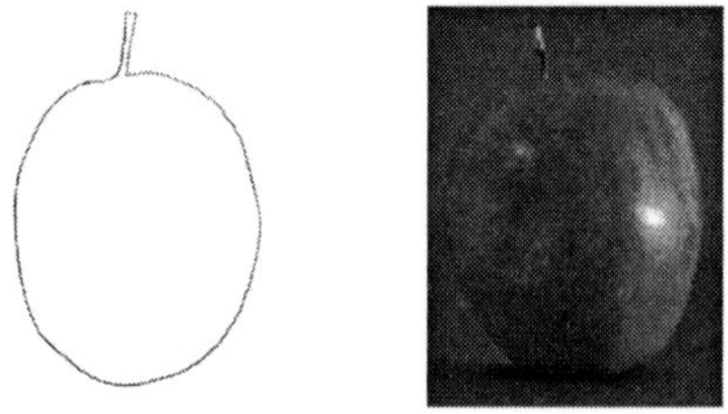

Figure 1.1: A sketch and an object with some similarities to the sketch

those properties which have been drawn intentionally from those which are accidental. There is little chance that two sketches which depict the same idea will be equal from a geometrical point of view, just as there is little chance that an imprecise sketch will equal geometrically a precise picture of a similar object (see Figure 1.1). Therefore, it is crucial to determine those characteristics which conceptually equal sketches have in common, i.e. shape information which has been drawn intentionally. Intentional shape properties are those which are necessary in order to specify an object, whereas accidental properties are unimportant for comparison purposes or even misleading. It is the aim of the current investigation to distinguish necessary from accidental shape properties, and to compare sketched objects by means of non-accidental shape properties.

We will get a first idea of how to approach this difficult topic by looking at the omnipresence of shape information in general, and especially at the importance of shape information in drawings. Discussing the special character of sketches, we will identify some problems which will allow us to state our objectives more precisely. These objectives will help us to deal with methodological issues.

1.1 Necessary versus accidental properties

In several areas *shapes* have been identified as an important and quite challenging issue, among others, in perception and art (Arnheim, 1978), design (Itten, 1963), computer science (Marr, 1982), and neurology (DiCarlo & Maunsell, 2000); from his psychological point of view Palmer (1999) emphasises the importance of shape information:

> *Of all the properties we perceive about objects, shape is probably the most important. Its significance derives from the fact that it is the most informative visible property in the following sense: Shape allows a perceiver to predict more facts about an object than any other property. Suppose, for example, you knew the color of an object but nothing else. What could you predict from the knowledge that it was, say, a certain shade of red? Most likely, almost nothing. But suppose you knew its shape and nothing else. From knowledge that it was the shape of, say,*

an apple, you would be able to predict a great deal about it, what it is used for, how it might taste, where it might be found, and so forth. In this sense, at least, shape is the single most significant property we perceive about objects.

We are accustomed to using shapes not only to predict facts about objects, but also simply to distinguish objects by means of their shapes; we use them to navigate around diversely formed objects, and to grasp, carry, and put away different objects. In language we use various adjectives which describe shape properties, such as *thin*, *elongated*, *round*, etc. Sometimes we even reason with the aid of shapes, for example, when planning a route which we have to follow. In short, our thoughts and actions are closely bound to shape.

Here we are especially interested in pictorial information, and we are looking for the distinctions that we make when dealing with shape information in sketches. Obviously, we are easily able to distinguish different shapes — for example, the letters of the present text, and even the letters of thousands of different handwriting styles. We want to investigate what the different spatial structures are which can be distinguished so effortlessly, and which are important in categorising different kinds of shapes.

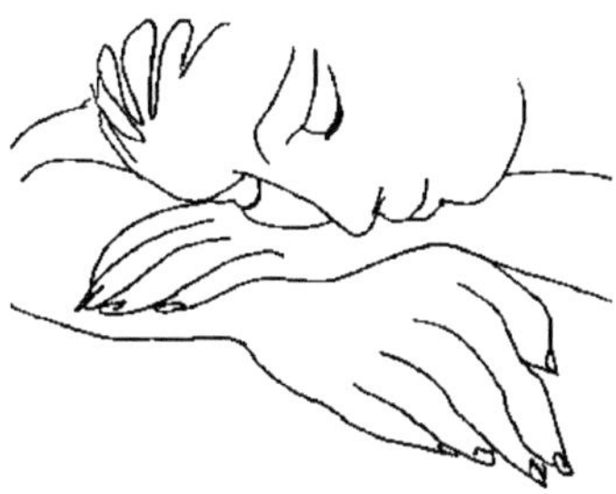

Figure 1.2: Drawing of a woman after a painting by Pablo Picasso: *War and Peace* (1952), Picasso Museum of Barcelona, Spain

The importance of shape information in drawings is a matter of particular interest for us. Artists, and especially caricaturists, are able to convey single objects or even complex scenes by merely drawing a few lines. The situation is different for coloured pictures. Colour, light, shadow, and everything that we see when observing natural scenes are helpful, even necessary, in categorising objects. But the richness of colour information does not function without shape, and actually such information induces us to see shapes. This may be the reason why we are able effortlessly to categorise the objects in a sketch which is made up simply of black lines on a white piece of paper, i.e. which is made up only of shape information, like the face in Figure 1.2. Such shapes sometimes contain only a few lines, so that we are faced with exceedingly stylised shapes which are confined to a minimum of spatial information — yet they are precise enough for us to categorise them. Obviously, some kinds of relationships in caricatures and stylised drawings are particularly relevant, so that our recognition abilities do not suffer from a lack of information in correctly categorising objects. These relationships can distinguish even categories of objects which would be highly complex in a naturalistic, non-stylised depiction, such as faces (as demonstrated above), or animals. An example of the latter is provided by Picasso's drawings of a bull at different levels of abstraction in Figure 1.3.

Here, we are not interested in recognising object categories by shape information, but in *comparing* images using shapes. We claim that the shape of any sketched object provides spatial structures which are useful in finding other images containing similarly shaped objects. These objects may actually belong to the same category as the query object, but they may also belong to other categories which just show similar spatial structures. Horses or dogs may display similar spatial structures to the bull depicted at the most abstract level in Figure 1.3. The specific properties of the bull get lost in an abstract depiction, but abstract levels are still good at pointing out spatial structures such as the four legs, the tail, and the approximate size and shape of the body. These spatial structures are necessary in order to determine some category. By contrast, specific properties supplement the necessary spatial structures in order to depict a particular object which is an instance of that category; the top-left bull, for example, looks less fierce than the bull to its right. The change in the depiction of the bull from a precise instance with a lot of specific properties to a depiction made up only of spatially necessary information is illustrated step by step in Figure 1.3.

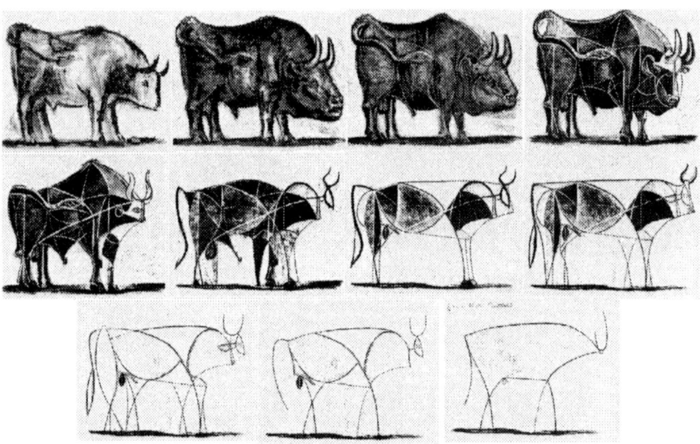

Figure 1.3: Drawings of a bull after paintings by Pablo Picasso: *Le Taureau* (1945/1946), The Museum of Modern Arts, New York, USA

Shape information, it follows, can be divided up into necessary relations between spatial structures that make up some category, and supplementary properties which vary in some details but maintain the overall appearance. This distinction between *necessary* and *supplementary* spatial structures corresponds to the distinction made at the beginning between *intentional* and *accidental* properties. Necessary properties are both intended and essential to categorising the drawing. Supplementary properties are inessential; drawn by mistake, or for decorative purposes, or to characterise a specific object instance. Which shape properties are drawn intentionally can be investigated from a psychological point of view. Such investigations need concepts which show what kinds of shape properties exist. We shall analyse the spectrum of shape properties and investigate how spatially necessary information can be characterised. That is, we shall investigate from the point of view of information theory what kinds of geometrical constraints apply at different levels of spatial precision in pictures, and we shall identify a number of distinctive features, all of which are candidates for being spatially necessary properties in sketches. Thus, we are faced with the problem of distinguishing two kinds of shape feature: necessary versus accidental properties, the former being crucial for comparing sketches.

> **Problem:** *Sketches comprise necessary and accidental properties which need to be distinguished.*

To make our point clear, we shall give an example. Someone is looking for specific objects in an image database. He has some specific characteristics in mind, which are satisfied by the objects he is looking for. Instead of studying the whole image database, which is irksome and will take a long time, he would like to sketch the desired objects. These sketches should then serve as graphical queries for a search engine for images.

The sketched objects are imprecise, i.e. shapes, positions, orientations, and sizes are indeterminate; they are accidentally drawn the way they are. One could imagine arbitrarily many different sketches, similar or even quite dissimilar to those shown in Figure 1.4, each one representing the same idea. What is necessary is the overall structure of the lines representing differently shaped objects, the left one being an ellipsoid which points upward and the other one being wide. The one on the right hand side has some bumps at the bottom, the left one not. The rough location of the stalks at the top matters, and probably that one points (fairly) straight upwards while the other one is bent. The determination of these necessary qualities is crucial.

1.2 Perceptually aided distinctions

We will get down to the problem of shape in sketches in the following way. A great deal of work has been based on metrical shape descriptions, which all have in common that they over-determine the properties of shapes in the following way. Given any sketch, metrical shape descriptions describe precisely the instance at hand, rather than the category to which that instance belongs.

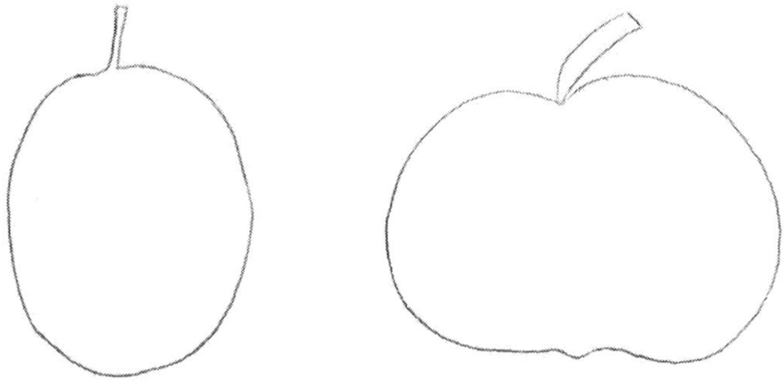

Figure 1.4: Two sketched query objects

This works well whenever there is no big difference between an instance and its category. But this does not apply to sketches, which will often be very different, imprecise drawings. Moreover, metrical shape descriptions reduce shape information to an absolute scale rather than directly addressing the qualities of a given sketch which characterise the category to which it belongs. A metrical representation can even be misleading, since it describes everything equally precisely including inessential details which may have been drawn by mistake. With a metrical representation, accidental properties have to be separated from non-accidental properties, which is quite a difficult procedure when all properties, accidental and non-accidental, have been represented equally precisely. Therefore, we shall focus on properties which are not based on metrical information, but which are distinguishable using perceptual systems which have no access to any metrical information. Instead of metrical distinctions, only some *perceivable distinctions* matter. But what do we mean exactly by perceivable distinctions? What kinds of such distinctions exist?

Let us, for example, consider the distinction between acute and obtuse angles: this distinction is easily perceivable, even when the angles are presented separately. We can distinguish mentally as well as visually between angles which are both acute, both obtuse, or different, because we are concerned with two simple shape categories which we can recall from memory. By contrast, the distinction between two slightly different acute angles is not so easily perceivable, and we have difficulties in imagining them. Mentally they melt down into one single acute angle, as if only one kind of acute angle existed. The reader is asked to imagine two acute angles, differing by less than 10°. How different do they look? We have to arrange these similar angles in a special way in order to distinguish them, for instance, by drawing them on top of each other, or next to each other on a piece of paper. We refer to the former category of distinctions, for example between acute and obtuse angles or between simple attributes such as round and angular, as *mentally aided distinctions.* Distinctions at this level are mentally aided, because these distinctions can be imagined mentally. We refer to the latter category of distinctions, for example between two similar acute angles arranged so that a direct comparison shows their difference, as *perceptually aided distinctions.* Distinctions at this level are perceptually aided, because they require comparison by perception to permit differentiation.

Perceptually aided distinctions are discussed by Freksa (1992b) who refers to them as to *qualitative knowledge*:

> *"... qualitative knowledge is obtained by comparing features within the object domain rather than by measuring them in terms of some artificial external scale. Thus, qualitative knowledge is relative knowledge where the reference entity is a single value rather than a whole set of categories. For example, if we compare two objects along a one-dimensional criterion, say length, we can come up with three possible qualitative judgements: the first object can be shorter (<), equal (=), or longer (>) in comparison with the second object. From a representation-theoretical point of view, a major difference between the two approaches is that measuring requires an intermediate domain in which the scale is defined while comparisons may be performed directly in the object domain." (Freksa, 1992b)*

Thus, according to Freksa, features can be either described using an external scale or compared directly to each other. In the latter case two acute angles, for instance, could be perceived, compared, and then judged as equal or unequal; in this case we would be faced with a binary qualitative judgement. Since such comparisons in an object domain can be made by means of perception alone, we are talking here more precisely about perceptually aided comparisons and, as a consequence, about perceptually aided distinctions. By contrast, we refer to those distinctions that require artificial external scales or an intermediate domain as *metrically aided distinctions.* For example, two acute angles could be arranged in such a way that we could not perceive them as different; in order to recognise the difference between them, an intermediate domain would be required, such as a scale defining angles to two decimal places. While metrically aided distinctions are objectively defined by an *artificial* scale, perceptually aided and mentally aided distinctions are dependent on the skills of the *natural* perceptual system which makes a judgement. Images are good at providing the means for perceptually aided distinctions. Objects drawn on a sheet of paper can be directly compared by perception without requiring external scales, but still allow finer distinctions than can be mentally imagined. This is actually the reason why sketches are made: they provide means for representing the richness of objects and ideas visually, where mental capacities are more limited. It follows that sketches are closely related to perceptually aided distinctions. We will show later on how some simple perceptually aided or mentally aided distinctions allow us to compose a vast number of different shapes. As these distinctions are perceptually more salient than metrically aided distinctions they form a possible means of comparing sketches:

Thesis: *Necessary properties can be described by perceptually aided and mentally aided distinctions. By contrast, accidental properties are in most instances only distinguishable by metrically aided distinctions.*

1.3 Methodological notes

Our object of investigation concerns pictorial information, and we want to investigate how a subclass of images, namely sketches, can be characterised for comparison purposes. To this end, our work is related to the field of *knowledge representation* in artificial intelligence; more precisely, to spatial representations, since we want to represent shape information in sketches. Furthermore, we define reasoning mechanisms in accordance with reasoning procedures devised in the field of *qualitative spatial reasoning* for the purpose of comparisons. In this context, we are concerned with both the exploitation of well-structured domains of spatial relations in images, and the exploitation of their algebraic properties.

Epistemological issues are confined to the assumption that the three-level class distinction mentioned in the previous section provides a plausible basis for dealing with different kinds of distinctions when comparing sketches. In establishing our objectives it has to be taken into consideration that we are not concerned with the processing of visually sensory information in order to obtain shape information. This is the concern of image processing, and of the biology and psychology of perception. Rather, the subject is how shape information can be represented, and how to reason about shapes using such representations. From this perspective, the present work mainly proposes an ontological basis for pictorial information in sketches.

The results of the present work can be put into a larger methodological context. One result will be a representation of shape information in sketches that provides a source for studies about concepts of intentional shape properties in sketches. It is a matter for cognitive studies to decide which of our geometrically-driven distinctions are actually intentional from a psychological point of view. Giving a formal theory of spatial concepts, we provide a set of spatial *primitives* and relations which form a basis for experiments. In this way, one removes the subjective element which is always present when selecting arbitrary spatial concepts for designing an empirical experiment about perception or sketching. Therefore, our approach exemplifies how information science yields computationally sound methods which may be used to serve the needs of the empirical sciences — one possible way for interdisciplinary collaboration to work.

1.4 Overview

First, the role of shape information in sketches is described, especially its imprecise character in this area. A number of sketching systems are reviewed, and we conclude that current approaches are too specific for our purposes. Qualitative spatial approaches are then surveyed, and are found to form an encouraging basis for sketch comparisons. A classification of spatial relations is used from which we learn that existent qualitative representations lack appropriate concepts to characterise pictorial information, especially shape information in sketches.

Our own representation is introduced in the third chapter. This approach is motivated by the problems which we have identified in the previous chapter; it facilitates the representation of shape information in images and reasoning about shape relations. The fourth chapter is devoted to polygons, which appropriately approximate shapes for the purpose of qualitative descriptions. Local and global shape properties are introduced and discussed.

Next, we show how we fill the gap which has been identified in the review of existing qualitative representations. It is shown how it is possible to deal with shape information when spatially querying for pictures, particularly in order to solve the problem of comparing linear objects in sketches.

We discuss how our approach can be integrated into existing geometrical representations which are complemented by our new one. The notion of *positional-contrast* is introduced, which characterises the new approach in the context of sketch comparisons. That is, we provide a qualitative spatial representation which characterises perceptually aided distinctions by positional-contrast and which forms a basis for distinctions in pictorial space in general.

Chapter 2

The imprecision of sketches

In the first chapter, we learned that sketches raise the problem of distinguishing necessary from accidental properties, necessary properties being crucial for the purpose of sketch comparisons. In our thesis we propose to recognise them using perceptually or mentally aided distinctions. The first step in achieving this is to get a general idea of what representations are based on such distinctions. Several representations of perceptually aided spatial relations have been proposed in the field of *qualitative spatial reasoning*. But before we come to those representations we need to work out what is relevant when comparing sketches. It is primarily the imprecision of sketches that makes them difficult to handle. Hence, it is helpful to provide a framework for distinguishing different levels of spatial precision. We are then able to figure out where present approaches lack appropriate concepts for dealing with imprecise shape information such as that found in sketches.

2.1 Characterising sketches

Sketches are primarily made up of shape information. Therefore, we have to clarify precisely what we mean by shape. The importance of shapes in sketches is then discussed. After that we will show that present systems operate on shapes in a rather constrained way. This is due to the inherent imprecision of sketches, as we shall see when studying the literature about peoples' ability to deal with spatial information in memory and sketches.

2.1.1 Shapes in sketches

In the context of the present work, by shape we refer to the way in which parts are put together in order to make up an object, i.e. an object has a more or less complex intrinsic structure which we refer to as its *shape*. The arrangement of a number of disconnected objects form a *pattern*. In the simplest case a shape is a closed *simply connected region*, but sometimes a shape consists of a more complex structure which could be regarded as a pattern of disconnected parts contained within an object. However, we are always able to outline the shape's boundary, at least roughly. The boundary can be defined as the object's exterior, i.e. something that connects the object with its surroundings. Hence, the boundary is particularly important since it defines how an object presents itself to its surroundings, that is, how it is perceived and how it relates to other objects. We refer to the shape of the boundary as the *contour* of the object.

While colour, texture, or different painting modalities may sometimes be used, many sketches are mainly comprised of linear features drawn by pen or pencil. Among other things, in geography such linear features correspond to objects like paths, roads, rivers, borders, or coast lines; for an urban planner, such linear entities may represent roads, plots of land, and parks or their respective boundaries; a designer may draw the outline of some objects, with some details which make up a pattern characteristic of that object. We focus on linear entities since non-linear entities such as regions can be described using linear entities, especially in sketches — linear entities being the chief ingredients in sketches, particularly when a sketch is drawn using a pen or pencil.

A sketch is made on a sheet of paper, and can be inspected at a glance, probably taking into account different spatial scales by looking at them at arm's-length as well as close-up. As a consequence, different spatial dimensions can be considered on equal terms: position, orientation, size, and shape can be inspected at different spatial scales, all these dimensions showing perceptually aided distinctions. This is quite different from the most prominent applications which deal with spatial information. In navigation, point-like abstractions of landmarks are useful for orientation purposes, and in geographical information systems cardinal directions between points matter, along with neighbourhood relationships between regions. Information about shape and frequently even size is unnecessary in these applications. But for our purposes shape and size become important. Sketches belong to the domain of pictures which have another ontological status, rather than that of navigation systems or geographical information systems. Pictures can be regarded as similar to an extreme of the types of information geographers deal with. Consider Tobler's First Law of Geography: *Everything is related to everything else, but near things are more related than those far apart* (Tobler, 1970). In geography the considered objects, cities for example, are far away, and it is sufficient to state that Bamberg is south of Hamburg regardless of the size and shape of these cities. By contrast, everything contained in a picture is so close to everything else that they relate in one way or another. In particular, shape information becomes important. We refer to the domain of pictures as *pictorial space*, in contrast to other kinds of space distinguished in spatial reasoning (cf. Freundschuh & Egenhofer, 1997); *vista space* is sometimes an appropriate view on problems in the context of navigation, as is *large scale space* in the context of geographical information systems.

Pictorial space requires, in equal measure, information about position, orientation, size, and shape, because pictorial space provides us simultaneously with all these types of information, all of them together determining the special character of a given sketch. Thus, sketches are a great challenge not only due to the need to distinguish between the necessary and accidental properties, but also in terms of integrating different spatial dimensions. This holds in particular when one is interested in describing not only single shapes but also the relations between differently shaped (and differently sized) objects, and their relative positions and orientations. We have to take into account all these dimensions, and the relationships between them. Additionally, sketches are made up of linear entities, and we conclude that we have to describe shapes (with regard to the mentioned spatial dimensions) using linear entities. Do linear entities actually play such a decisive role in existing sketching systems?

2.1.2 Electronic sketching

Yes, they do. Existing electronic sketching systems predominantly use pens as input devices, allowing the user to sketch objects using linear components. These systems, are intended to provide user interfaces which feel as natural as pencil and paper, but which also provide intelligent means for, among other things, automatically recognising drawn objects. It is often argued that it is an important property of sketches that they look informal (Gross, 1998), (Landay, 1998). When working out new ideas, people are inclined to sketch their thoughts informally. Informal depictions invite modification and further brainstorming, whereas precise graphics look neat and perfect, suggesting a finished design. But transferring *informal sketches* to the computer, is often a complicated and tedious process (Stahovich, 1998). The same holds for graphical queries in the context of image databases. A sketched query should look informal, too. A neat and perfect graphical query would suggest that the user has a perfectly elaborated idea in mind. But frequently what the user pictures (and what he knows about what he intends to find) is rather more coarse and vague than this.

Those aspects of the problem of intelligent user interfaces which bridge the gap between informal sketches and neat graphics form the primary context in which sketching is investigated in applied computer science and intelligent systems. In this context, we are only interested in one aspect, namely how these systems deal with shapes. For this purpose, let us first have a look at how these systems bypass two of the most difficult problems in computer vision, namely *segmentation* and *object recognition*. In the spatial query by sketch system of (Egenhofer, 1997) the user has to select first the object category of the object he intends to draw; examples are rivers, woods, or land parcels. By this means, the problem of segregating and grouping elements is bypassed. After the user has drawn the desired configuration, the system automatically translates that sketch into a representation that can be processed against a geographic database.

By contrast to (Egenhofer, 1997), most systems, such as the one from (Alvarado & Davis, 2001), release the user from selecting an object category first. Such systems make use of heuristics about how people draw. For example, they assume that people tend to draw all of one object before moving to a new one, that is, they assume that consecutive drawn strokes are likely to belong to one single object. As soon as a number of consecutive strokes fit the template of a given shape category, that object is displayed differently coloured in order to indicate what has been recognised by the system. The user can then make corrections directly if necessary.

(Alvarado & Davis, 2001) use templates of predefined shapes defined by basic geometry, for instance, a circle template which matches if all the points on a drawn stroke lie at roughly the same distance from the average coordinate of that stroke. (Hammond & Davis, 2003) arranges these predefined shapes in an inheritance hierarchy, complex shapes being made up of primitives, to which *point*, *path*, *line*, *Beziercurve*, and *spiral* belong, all being derived from an abstract concept *shape* that provides a number of components and properties, such as bounding-box, centre-point, width, and height. (Sezgin, Stahovich & Davis, 2001) makes use of speed data to allow the detection of vertices not only by curvature maxima but also by looking for points along drawn strokes with minima of speed. In (Landay & Myers, 2001) four primitive components can be recognised, namely rectangles, squiggly lines, straight lines, and ellipses which combine to form seven basic widgets as well as combinations of these in the context of user interface design. Moreover, such combinations are described in terms of the spatial relationships which exist between them, that is, whether components *contain* other components, whether a component is *left*, *right*, *above*, or *below* another one, or whether a component is in a *vertical* or *horizontal* sequence of any combination of components. (Hse, Shilman & Newton, 2004) fragment a variety of sketched symbols like squares, ovals, trapezoids, or pentagons into simpler structures which are approximated by line segments and elliptical arcs. This fragmentation can then be used for template matching in order to recognise symbols. (Skubic, Blisard, Bailey & Matsakis, 2004) translate sketched route maps into linguistic descriptions. Different objects as well as the route are separated by delimiters during the sketching process. The relative orientation between two objects is determined by the histogram of forces, as introduced by (Matsakis & Wendling, 1999), and the orientation is discretised into one of sixteen possible directions. The route is described from an egocentric point of view taking into account the route's changes in orientation and the orientations of the drawn landmark objects.

Those approaches discussed so far optimise interaction naturalness at the cost of tightly restricted domains. By contrast, (Ferguson & Forbus, 2002) can be used in many different domains at the cost of reduced interaction naturalness. They call a collection of strokes a glyph representing an entity or relationship, and avoid object recognition either by selecting the interpretation of a glyph explicitly from menus, or by identifying the glyph by spoken words which the system attaches to the drawn glyph in a multimodal environment. The glyph itself is not further analysed. Meaningful spatial relationships between glyph symbols are expressed by arrows, for example, in course-of-action diagrams, in biological sequences, or in structural descriptions. For this purpose, topological relationships as described by RCC8 (Randell, Cui & Cohn, 1992) are used, as well as some simple orientation relationships, such as *left of*, *right of*, *above*, and *below*. We have seen that present approaches mainly treat shapes metrically, with the exception of (Ferguson & Forbus, 2002) and (Skubic et al., 2004) who use ordinal information in order to describe the relative orientation between objects. Besides orientation information (Landay & Myers, 2001) also use simple topological relations, such as *contains*, and (Ferguson & Forbus, 2002) make use of RCC8.

We come to the conclusion that electronic sketching systems do not address the distinction of necessary and accidental properties. They do not even make any clear distinction between the process of *obtaining* and *representing* shapes, let alone attempt any reasoning procedures about such representations, particularly in order to deal with incomplete and imprecise shapes, which is quite important, given that sketches are imprecise by nature. This imprecision is either treated by confining sketching systems to small, well defined sets of shapes, or it is left to the user to tell the system directly what he intended to draw. Is there a way in which we can treat imprecision in sketches more generally?

2.1.3 The imprecision of sketches

What, in fact, do we mean by imprecision? In what sense are sketches imprecise? Imprecision is related to the question of what kinds of distinction become more difficult (or impossible) as something gets more imprecise. In the context of sketching, details of objects and the relationships between them are imprecise. On the other hand, there are at least some coarse object properties and relationships which are necessary for any sketch to make sense. These distinctions are related to what people memorise, what they are able to recall from memory, and what they are able to sketch.

A number of investigations had been carried out concerning the kinds of spatial knowledge people acquire about their environments and which they use when asked to remember environments, describe a route, sketch a map, or make judgments about locations, directions, and distances. These investigations have brought forth much evidence that memory and judgment are systematically distorted. For example, people sometimes make errors in judging the direction from one city to another. Such errors are attributed to hierarchical representations of space. If people do not remember the absolute locations of cities, they infer the relative locations of cities from the locations of their superset states. Boundaries of regions are sometimes quite complex so the rough direction from a point in one state to a point in another is not always the same for arbitrary pairs of points (Stevens & Coupe, 1978). Students were asked to estimate distances between buildings in a city. These buildings were grouped according to function, commercial or educational. Distances between buildings belonging to different groups were overestimated relative to distances between buildings of the same category (Hirtle & Jonides, 1985). Similarly, in another experiment students were asked to estimate distances between cities. Distances between nearby cities were overestimated relative to pairs of cities being far away (Holyoak & Mah, 1982). Less well-known locations are described with respect to more familiar ones, and people also seem to remember them that way. In the same way, we normally describe situations from our own perspective rather than from any other point of view (Couclelis, Golledge, Gale & Tobler, 1987). Remembering less well-known locations relative to landmarks induces asymmetric distance judgements. People judge distances from ordinary buildings to a landmark to be smaller than the distance the other way round (Sadalla, Burroughs & Staplin, 1980). Nearly aligned locations tend to be grouped in memory, and then remembered as more closely aligned than they actually are. Similarly, locations remembered with respect to a frame of reference can lead to rotation distortions. Students who had to place cutouts of South America correctly placed it upright, while it actually appears to be tilted on one side in a north-south east-west frame (Tversky, 1981). Irregular geographic features are sometimes regularised. Parisians straighten out the Seine (Milgram & Jodelet, 1976), and Americans straighten out the Canadian border (Stevens & Coupe, 1978), (Tversky, 1992). Turns and angles are regularised to right angles (Byrne, 1979). Distances are judged longer when a route has barriers or detours (Cohen, Baldwin & Sherman, 1978), when a route has more turns or nodes (Sadalla & Magel, 1980), and when a route has more clutter (Thorndyke, 1981).

Besides this evidence about distortions in large scale space, perceptual psychologists have accumulated evidence showing distortions in pictorial space. For example, depending on the context, lines which are equal in length are judged to be different in length, as shown by the Müller-Lyer illusion or Ponzo figure; vertical lines are overestimated with regard to horizontal lines (Figure 2.1). Lines may even be perceived where there are actually no lines in the stimulus, as in the case of the Kanizsa triangle (cf. Gregory (1997)).

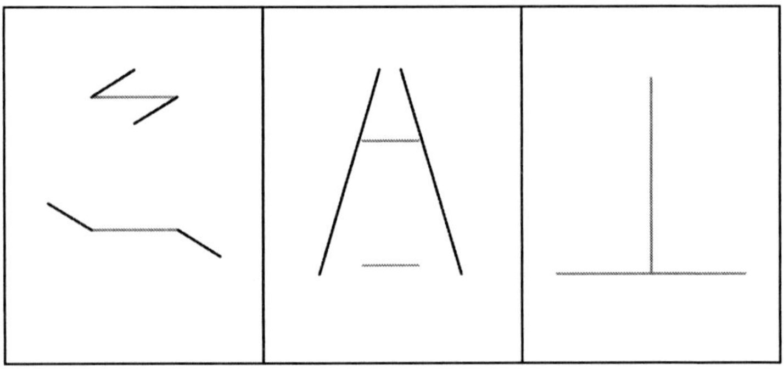

Figure 2.1: Visual illusions: variation of the Müller-Lyer illusion (left); Ponzo figure (middle); horizontal-vertical illusion (right) — compare the grey lines

Many investigations concerning perceptual processes have been carried out, but comparatively little is known about the sketching behaviour of humans. Frequently, sketching is used in experiments in order to test the abilities of people with brain lesions and other disabilities. For instance, humans who suffer from shape-agnosticism are not able to draw a simple shape such as the shape of the letter S (Goldenberg, 1998). Instead, only fragments of the shape are drawn. People who suffer from other diseases show in their drawings particular distortions and incomplete shapes. Studies involving blind people who are able to draw by making raised-line drawings are quite exciting, suggesting that the development of drawing in blind and sighted children may be similar because haptics provides access to many of the same spatial principles as vision (Kennedy, 2003).

There are also various studies which show characteristic distortions and limitations of people who draw something and who do not suffer from specific disabilities. Many people are unable to match the apparent directions of objects in a perspective picture of a scene, even in simple copying tasks (Willats, 1997). Inconsistency between different ways of showing depth is the norm in drawings by children and adults (Milbrath, 1998), and inconsistencies between overlaps and possible projective schemes also occur(see (Willats, 1997) and (Landerer, 2000)). Additionally, (Tversky, Zacks, Lee & Heiser, 2000) demonstrated that people use small sets of schematic figures to convey certain context specific concepts: in sketch maps, blobs, straight lines, curved lines, and crosses are used systematically to convey information about geographical features; in graphs, bars indicate discrete comparisons while lines indicate trends; in mechanical diagrams, arrows signify order of functional operations.

The method known as *serial reproduction* is mainly used in the linguistic analysis of text understanding (Stadler & Wildgen, 1987) but has also been applied in the redrawing of simple pictures (Bartlett, 1932). Especially, there are a number of studies in the context of *dot displacements* in which a piece of paper with a single dot on it is exposed for a short time (less than one second) to a subject who, after some seconds, has to reproduce the localisation of the dot on another piece of paper. This reproduction is then shown to another subject who is given the same task. The result is presented to the next subject, and so on. The superposition of the reproductions shows a characteristic wandering of the point which seems to be determined by the shape of the paper used. For example, taking a DIN A4 paper the point wanders from the middle of the paper, where the dot is placed at the beginning, to one of the corners of the sheet of paper (Stadler, Richter & Pfaff, 1991). Regularities of the processes which are shown by such experiments are interpreted in Gestalt psychological terms as an effect of the tendency towards *Prägnanz*. Distortions are shown when reproducing something by drawing, and these distortions are very similar even for different subjects (Stadler et al., 1991).

We have learned that human beings distort spatial information in memory as well as with regard to pictorial space. From this, it follows that sketched ideas which have been recalled from memory are imprecise, although needless to say additional imprecision is typically introduced in the sketching, since most people do not sketch perfectly. Therefore, precise measurements and precise comparisons of sketches are not meaningful. But the previous evidence also shows that humans tend to distort spatial information in certain ways, which are independent of subject. For this reason, it is meaningful to compare sketches at a coarse granularity level of spatial information where subject-independent tendencies are valid. In the following we shall see how it is possible to obtain such a coarse granularity level.

2.1.4 Straightening away imprecision

Having studied perceptual processes from the point of view of information theory Attneave (1954) came to the conclusion that our perceptual systems are confronted with much redundancy. He proposed a technique for obtaining coarser shape descriptions which are less redundant, but which nevertheless maintain perceptual distinctions. This technique could be conceived as a means of reducing noisy information, or, as we referred to it before, reducing accidental shape properties. It was devised by analysing the results of the following experiment.

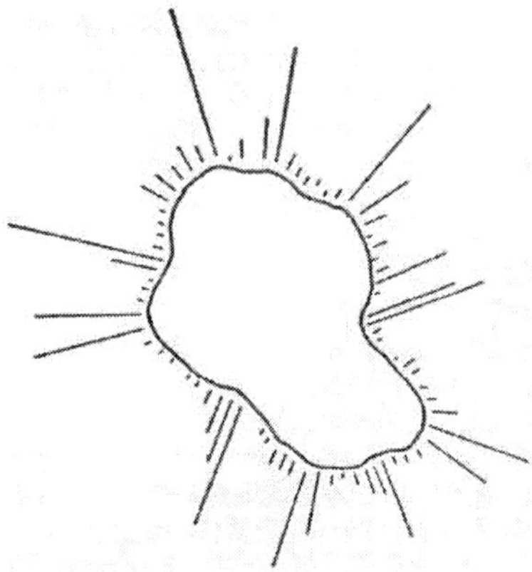

Figure 2.2: Subjects attempted to approximate the closed figure with a pattern of 10 dots. Radiating bars indicate the relative frequency with which various portions of the outline were represented by dots chosen — Figure 2 in Attneave (1954)

Eighty subjects were instructed to draw, for each of 16 outline shapes, a pattern of 10 dots which would resemble the shape as closely as possible, and then to indicate on the original outline the exact places which the dots represented. A sample of the results is shown in Figure 2.2: radial bars indicate the relative frequency with which dots were placed on each of the segments into which the contour was divided for scoring purposes. It became clear that subjects show a great deal of agreement in their abstractions of points best representing the shape, and most of these points are taken from regions where the contour is most different from a straight line, i.e. where local curvature maxima in the form of concavities or convexities are. This conclusion was verified by detailed comparisons of dot frequencies with measured curvatures on both the figure shown and others. These results have recently been confirmed by (Norman, Phillips & Ross, 2001).

As a consequence Attneave concluded that the most salient parts for the perception of shapes (that is, the most informative perceptually) are the points of maximum curvature along an objects's contour. Figure 2.3, usually referred to as *Attneave's cat*, was created (to back up his assertions) by taking the points of maximal curvature from the contour of a cat, and then connecting them with straight lines. The resulting figure is easily recognisable as depicting a cat, and this suggests that an economical encoding of two-dimensional shapes could be obtained from identifying the points of maximal curvature and connecting them in this fashion. This is how polygons come into play.

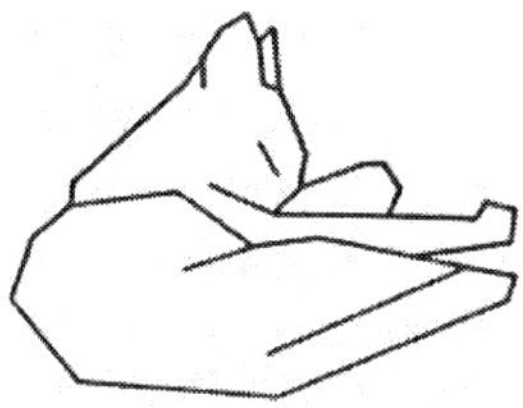

Figure 2.3: Attneave's cat, Fred Attneave (1954)

Attneave's technique has been widely used, and, from the technical point of view, polygons have the advantage that there are already many methods for generating polygonal approximations. For instance, (Cordella & Vento, 2000) reviewed almost one hundred papers in the field of computer vision, and found that 58% of the reviewed systems applied thinning techniques and polygonal approximations to shapes. Polygonal approximation algorithms range from simple ones, like (Douglas & Peucker, 1973), to more sophisticated algorithms (Horng & Li, 2002) which aim to find optimal polygonal approximations. (Rosin, 2003) even developed several measures to assess the stability of such algorithms.

Hence, coarser granularity levels of sketches can be obtained by approximating the linear curves of a sketch using polygons. This has the advantage that a number of different granularity levels can be taken into account — at least one of which should be appropriate depending on the imprecision in hand. Particularly, we are not interested in taking into account the details of a drawing which has been scrawled down with a shaky hand, and even a well trained artist is not capable of drawing, for instance, a perfectly straight line or a perfect arc. The polygonal approximation of an (inherently somewhat inaccurate) drawn figure, forms an appropriate basis for comparing it with others since such a polygonal approximation is restricted to depicting perceptually aided distinctions.

Regardless of exactly what can be drawn to what degree of accuracy by human beings, we will now investigate what different levels of spatial precision can be defined, before moving on to identify a level of spatial precision which allows us to deal with imprecision in sketches more generally than those systems which we have been reviewing so far. This level of spatial precision will then be used for characterising polygons.

2.2 Imprecise spatial information

We are able to compare sketches, when looking at them, even though they are imprecise, as a result of the way we observe features and relationships among them. In this section we intend to examine what types of features and relations support meaningful comparisons of imprecise sketches. The previous sections taught us that in the context of sketches meaningful comparisons are to be made at coarse granularity levels, at which we are able to focus on perceptually aided distinctions. Thus, we shall analyse representations in terms of their sensitivity to imprecise spatial information, and we aim to identify a representation which is both coarse enough to deal with accidental properties and fine enough to represent all necessary shape features.

2.2.1 Levels of spatial precision

We shall classify spatial representations accordingly to their sensitivity to imprecise information, placing at the lowest level those representations which are most sensitive, i.e. all conceivable transformations change the properties of shapes which are represented at this level of spatial precision. By contrast, at the most insensitive level many transformations can be applied to a shape without changing those properties which are defined by a representation at this level.

Examples of representations which are quite insensitive to imprecision are topological representations. Such representations distinguish, for instance, regions having a hole from those having not a hole, or *self-connected regions* from regions which are separated into multiple pieces, as well as relations between regions, such as disconnected regions or overlapping regions. We refer to this level as the level of topology. Topological distinctions are quite robust with respect to imprecision since a region with a hole, for instance, remains a region with a hole regardless of the appearance of either the hole or its surrounding region.

In addition to the existence of topological features or relations one might also consider the arrangement of elements. Examples of this include the linear arrangement of three objects, one being exactly in between the other two, and the cyclic ordering of points indicating time on a watch, along with orientation information such as the four points of the compass. This level is referred to as the level of ordinal information. Small changes of the positions of elements will often not entail changes in the ordering of those elements, i.e. the level of ordinal information is still quite robust. But it is less robust than the level of topology since a change in the ordering of objects will not affect topology. Projections, such as orthogonal projections or central projections, belong to this level of spatial precision since they consider only the ordering of elements between the original shape and the projected one. Therefore, we can also refer to this level of spatial precision as the level of projections.

The next level is that of affine transformations. The concept of congruence is a specific affine transformation. When we are looking at two congruent figures, there are sides and angles that match up in each figure. These sides and angles are called corresponding sides and corresponding angles, and these corresponding parts are also referred to as congruent. This means that corresponding sides are equal in length and that corresponding angles are equal in degrees. That is, congruent objects are exactly the same size and shape. From the point of view of sketching it is certainly difficult to produce duplicates of an object. Rather than taking sides of equal length, a weaker demand is that ratios of corresponding sides have to be in proportion, as in the case of similarity mappings which define another specific case of affine transformations. Architects and map-makers, for instance, use such scale drawings — the maintenance of ratios being particularly important in these fields. Reflections, translations, and dilations also belong to this level of spatial precision since all these transformations maintain complete shapes and change either only the positions and orientations of shapes, or else only their size.

Finally, any representation which considers absolute values is exceedingly sensitive. For example, taking positions of points using a Cartesian reference system, arbitrarily small changes lead to changes of the recorded position. What distinguishes this level is that comparisons are made with regard to some artificial external scale while at the first three levels objects or their properties are compared directly with each other. Further levels are conceivable at which other properties are considered important. But these four levels provide a framework relative to which representations can be compared for the purpose of finding a representation for imprecise sketches.

Specific transformations can be applied to any shape without changing it with respect to each specific level, except at the absolute level, where every transformation changes the shape and even the smallest irregularities or variations lead to differences between shapes. On the level of affine transformations, in particular in the case of similarity mappings, shapes can be scaled up or down without affecting their equality. But correct ratios are difficult to render when sketching. Hence, the first two levels seem to be inappropriate for representing sketches. At the ordinal level, i.e. the level of projections, distances can be also changed. But the arrangement of elements and their respective orientations are not allowed to change. The order of elements is definitely not difficult to maintain, even in an imprecise sketch. As such, representations at this level are probably useful for representing sketched objects. At the topological level, all topological transformations can be applied to the shape. This allows quite large changes, for instance, two closed regions are topologically equivalent even though they may have very different shapes. Being robust against specific changes each level can be described by a number of invariants which characterise that level. Some of the invariants which apply at each level are shown in Table 2.1.

As an example, let us consider the rectangle on the left hand side of Figure 2.4. At each level different changes are applied to the rectangle, while maintaining equality at this level. The absolute level does not allow any changes, or this specific rectangle will be changed. At the level of affine transformations, (in particular considering similarity mappings), variations are generated by scaling the rectangle up and down. At the ordinal (projection) level, the ordering of corners is considered, and two corners can, for example, be related by directions, such

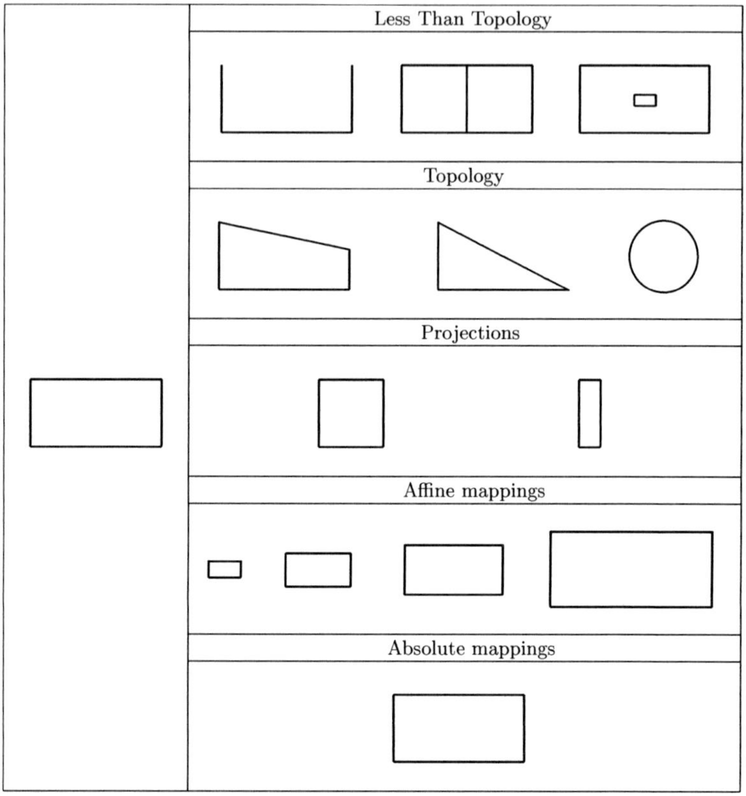

Figure 2.4: Equality at different spatial levels

Table 2.1: Different levels of precision correspond to different sets of invariants

Level	Invariants
Topology	Ordering & Position & Ratio & Scale
Projections	Position & Ratio & Scale
Affine mappings	Position & Scale
Absolute mappings	Nothing

as *below* or *left of*. At the topological level, any topological transformation can be applied to the rectangle and, for example, any quadrilateral, triangle, and circle are regarded as equivalent to it at this level. If every conceivable transformation is allowed (slicing through a contour, for instance), the rectangle can be distorted arbitrarily. The example demonstrates that at the topological level (and even more at the unrestricted level) the similarity rapidly decreases. For our purpose, the level of ordinal information seems to be the best candidate. We shall verify this in the following paragraph by discussing how existing approaches to qualitative spatial representations fit into this hierarchy of spatial precision. Moreover, at each level arbitrary variations in the resolution used are possible. For example, at the ordinal level one could choose between a resolution of 90° and 180° angles for the purpose of describing directions, or at the absolute level one could use an arbitrarily fine scale. Finally, we will have to decide both the level of spatial precision and the accuracy to be used at this level.

2.2.2 Qualitative spatial representations

How are the four levels of spatial precision we have defined related to shape properties? Studying the literature about visual shape perception, we learn that shape properties usually refer to general characteristics such as *round* versus *angular*, *symmetric* versus *asymmetric*, *regular* versus *irregular*, or *simple* versus *complex* (cf. Palmer, 1999). Although such qualities are used in several ways, these properties are quite general and only capable of coarsely classifying different kinds of shape. Looking at the variety of shapes in nature, it is amazing that we come up with only a few general characteristics. It is as if we lack the right vocabulary to describe the great variety of shapes in nature — there seems to be a large gap between perception and language. Therefore, we are interested in investigating whether there are perceptually aided shape characterisations other than those just mentioned.

Representations of spatial relations which are perceptually aided have been proposed in the field of qualitative spatial reasoning. In contrast to the general shape properties just mentioned, qualitative spatial reasoning approaches are distinguished by representations on the basis of which sound mathematical reasoning procedures are defined. Furthermore, the qualitative representations usually define sets of jointly exhaustive and pairwise disjoint relations, describing a particular dimension, such as directional information or topological relationships. A set of jointly exhaustive relations allow for all possibilities, i.e. no possible relation within the considered dimension is left undefined, and a set of pairwise disjoint relations avoids any ambiguities since exactly one relation holds for a given arrangement of objects.

Concepts in the field of qualitative spatial reasoning are restricted to rather small sets of relations representing distinctions important in the chosen context. Accordingly, we have to take into account what kinds of distinction are important in the context of imprecise sketches. Moreover, qualitative concepts are often driven by easily perceived distinctions; as we have already learned perceptual distinctions are important for the purpose of comparing sketches. Even more important, unlike image processing and computational geometry, which both deal in a highly precise way with spatial information, qualitative concepts are especially well suited to coping with imprecise and incomplete information — particularly important in the case of sketches. Therefore, it is worthwhile to examine approaches which describe shapes using qualitative spatial representations.

A general overview of qualitative reasoning approaches is provided by Cohn and Hazarika (2001). By contrast, we review those approaches with the emphasis being on what is relevant for characterising necessary shape properties of sketches. Given the requirements we have established, only those approaches which are related to shapes (and which are useful in properly deriving our own approach later on) are reviewed. At the absolute level, precise measurements are significant, that is, at this level we leave the domain of qualitative spatial representations. On the assumption that crucial characteristics of sketches are represented by qualitative spatial concepts, any approach at this level is outside the scope of our interest. Having also identified in section 2.2.1 the level of affine transformations as inadequate for the representation of sketches, the two remaining levels are those of topology and ordinal information (or projections). We will discuss representations at these levels in that order.

Egenhofer et al. (1991), Randell et al. (1992)

Topological relations for the purpose of spatial reasoning have been proposed by Egenhofer et al. (1991) and by Randell et al. (1992). They define different dyadic relations between extended regions. In their *Region Connection Calculus* (RCC), Randell, Cui, and Cohn (1992) have chosen an axiomatic formalisation based on the binary predicate $C(x, y)$, which indicates whether two regions x and y are connected or not. Egenhofer and Franzosa (1991) derived their formalisation by considering the intersections of boundaries and interiors for two regions. This approach is therefore called the 9-intersection calculus. Both approaches identify topological relations between two regions, such as those depicted in Figure 2.5.

Further relations are proposed by Egenhofer (1997) who takes the following extensions to topological relations (which he refers to as *detailed topological relations*, which are depicted in Figure 2.6) into account: the *boundedness* de-

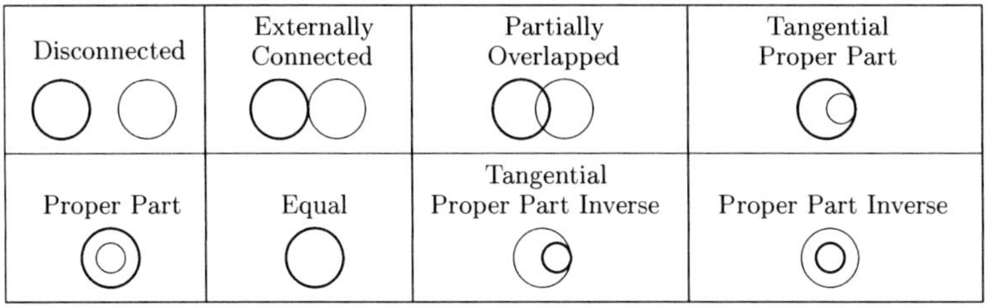

Figure 2.5: Eight topological relationships between two regions

scribes whether a one-dimensional boundary component shared by two regions is between them or on the same side of both (Figure 2.6.e); the *complement relationship* refers to a component which is next to either an open or closed exterior (Figure 2.6.f); the *dimension* of components is zero-dimensional if two regions intersect at a point or one-dimensional if they intersect along lines (Figure 2.6.b). Some of these *detailed topological relations* seem rather like ordinal relations: The *sequence* of components along a regions's boundary — such a sequence actually describing the ordering of those components (Figure 2.6.a); the *type* of component intersections distinguishing whether the boundary enters and leaves the component intersection from the same part or from different parts — again, this concerns the ordering of the objects' components (Figure 2.6.c); the *crossing direction* describing whether the boundary component leads into or out of the interior of the other region — which is obviously related to orientation information (Figure 2.6.d). Furthermore, Egenhofer uses cardinal directions which are defined by a bounding rectangle of a region partitioning space into nine areas. It is then possible to describe the location of another region in terms of the cardinal directions defined by the former region.

Especially interesting in the context of sketches is the description of relations between line segments. Egenhofer and Herring (1991) consider all possible relationships between two lines defined by the 9-intersection calculus. Unfortunately, intersection relations among arbitrarily curved lines are considered which may have simple or quite complex shapes. For this reason, visually different line relations fall into the same equivalence classes.

Topological distinctions are quite general. Primarily, these relations have been proposed for relating geographical objects. Topological relations are interesting in the context of sketching since they can readily be distinguished in a drawing (see the different relations in Figure 2.5). In fact, an approach for describing shapes at the topological level exists.

Cohn (1995)

(Cohn, 1995) proposes a topological approach which allows different concave shapes to be distinguished. This approach uses two primitive notions: that

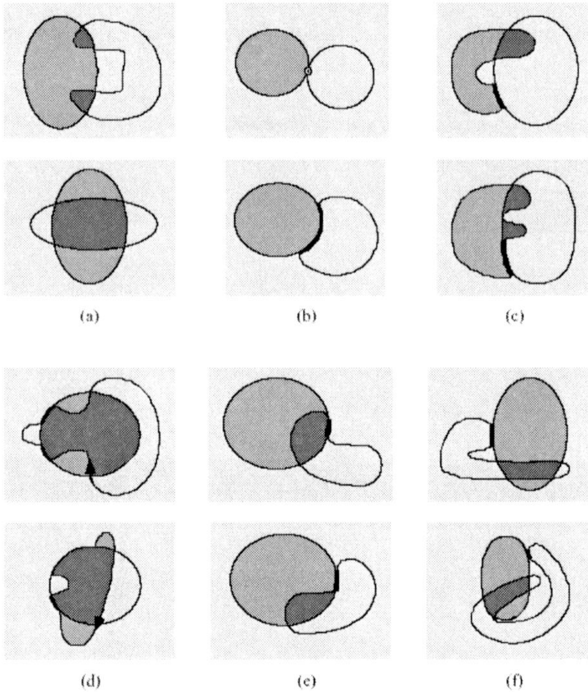

Figure 2.6: Detailed topological relations from Egenhofer (1997)

of two regions connecting, and that of the convex hull of a region. Once one takes the convex hull of a region, relationships between the shape itself and the different components of the difference between the shape and its convex hull can be described. The left hand side of Figure 2.7 shows an example. A refinement to this technique exploits the idea of recursive shape description in order to describe any non-convex component of the difference between the convex hull and the shape, as demonstrated by the example on the right hand side of Figure 2.7.

This approach is capable of describing different kinds of concave shape, at least coarsely. But different convex shapes, objects of different sizes, and patterns consisting of a number of objects cannot be described using it. More generally, no kind of topological shape description preserves angles, lengths, or ratios. As such, shapes of the kinds depicted in Figure 2.8 are topologically equivalent. This example shows that constraints imposed at the topological level are too weak, even for characterising necessary shape properties in sketches — distinctions at that level are easily made in sketches, but we shall see that there

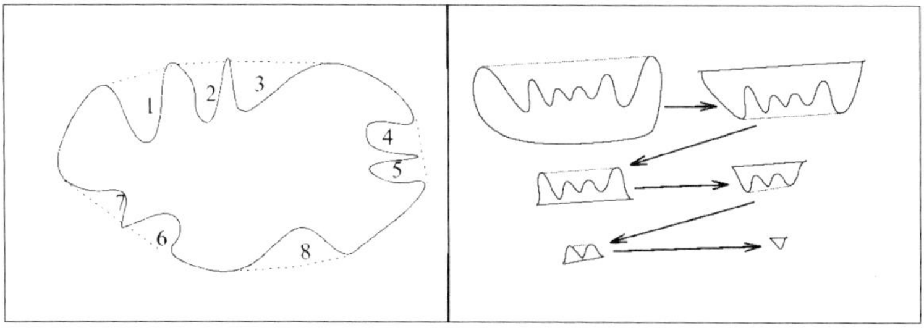

Figure 2.7: Left: A shape and its convex hull; Right: Hierarchical shape description — from Cohn (1995)

are further relations which are more precise than distinctions at the topological level, which still can be sketched readily. The ordering of some objects along a straight line, for example, is definitely as simple to preserve in drawings as topological relations are.

At the level of ordinal information approaches describe relative directions or the ordering of objects — two different ideas which are closely related (Schlieder, 1995b). However, they allow us to describe shapes more precisely than at the topological level, which is probably the reason why there are more shape descriptions at this level. Whereas mainly regions, or connected point sets, are used at the topological level, at the level of ordinal information directions become important. Relations involving directions can be defined for the most simple entities, namely points. For this reason, most approaches at the level of ordinal information are defined using points, abstracting away information about size and shape (both being unimportant for various tasks in, for instance, geographical information systems). We shall see later on how such approaches form a useful basis on which to define shapes.

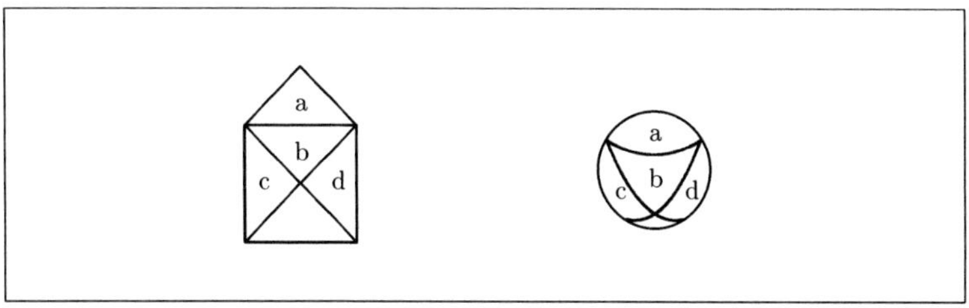

Figure 2.8: Two topologically equivalent, but rather different-looking figures; corresponding regions are labelled by the same letter

before	<
meets	*m*
overlaps	*o*
starts	*s*
finishes	*f*
during	*d*
equals	=
contains	*di*
is finished by	*fi*
is started by	*si*
is overlapped by	*oi*
is met by	*mi*
after	>

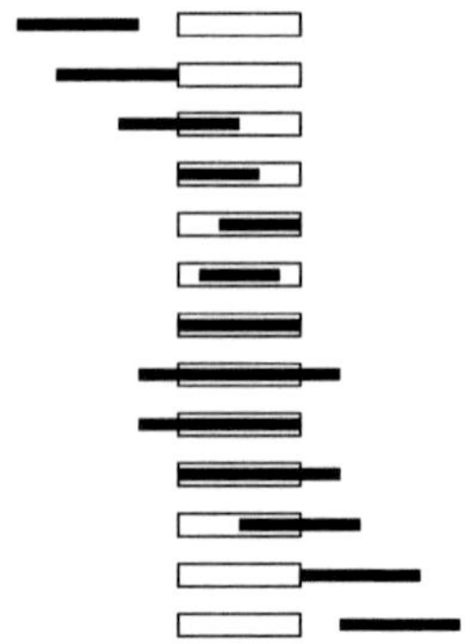

Figure 2.9: Allen's thirteen one-dimensional interval relations

Allen (1983), Freksa (1992)

Allen (1983) introduced a description of ordinal relations in one-dimension in terms of time intervals. There are thirteen such interval relations on the one-dimensional axis, which are depicted in Figure 3.2. Each relation can be defined in terms of the relative position between the four endpoints of two intervals. These interval relations are of particular interest to us since they are obviously perceptually aided. We will therefore show below what generalisations of this approach to two dimensions exist.

The interrelationships of Allen's interval relations have been investigated by Freksa (1992a), who arranges these relations in their *neighbourhood graph*. In this graph two relations are connected if they can be transformed into each other without meeting any other relation. Figure 2.10 shows the neighbourhood graph. Depending on the types of deformation of intervals and their relations allowed, different neighbourhood structures are obtained. If we fix three of the four endpoints and allow the fourth to be moved we obtain the A-neighbour relation. If we leave the length of intervals fixed and allow complete intervals to be moved, we obtain the B-neighbour relation. If we leave the location of intervals fixed and allow their lengths to vary, we obtain the C-neighbour relation. This concept of neighbourhoods is quite powerful, and allows to structure the domain of several qualitative representations, even of topological relations, with each neighbourhood structure defining an ordering of a set of relations.

Guesgen (1989), Hernandez (1992)

Guesgen (1989) was the first one who applied Allen's approach to the spatial domain. He argues that the definition of one-dimensional spatial relations is straightforward since there is a direct isomorphism between structures of time and structures of one-dimensional space. For the one-dimensional case Guesgen

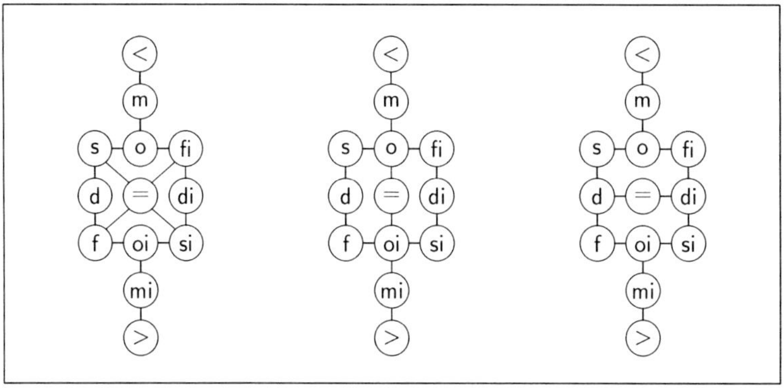

Figure 2.10: The thirteen Allen relations arranged in conceptual neighbourhood structures. Left: A-Neighbours, Middle: B-Neighbours, Right: C-Neighbours

distinguishes eight possible relations between two objects O_1 and O_2, namely O_1 left of O_2, O_1 attached to O_2, O_1 overlapping O_2, O_1 inside O_2, and their converse relations. For more than one dimension, Guesgen treats each spatial dimension independently in the same way as the one-dimensional case, with different dimensions arranged orthogonally. As a consequence, only rectangular objects which are aligned with the underlying absolute frame of reference can be adequately represented.

Hernandez (1992) combines topological relations and orientation relations. A point-like reference object determines eight orientation relations, and a primary object is described with respect to that reference object. Figure 2.11 shows the structure which integrates both topological relations and orientation relations from the viewpoint of one reference object. The d corresponds to the topological relation of the outmost ring, meaning *disjoint*; t describes relations at the medial ring, meaning *tangent*; o means *overlaps*, c *containment*, i *included*, $i@b$ *included − at − boarder*, etc.; for instance, $[d, l]$ means $during - left$ and $[o, rf]$ means $overlaps - right - front$. Both Guesgen and Hernandez base their representation on a static reference system.

Freksa et al. (1992)

Instead of using an absolute frame of reference the approach of Freksa and Zimmermann (1992), (Freksa, 1992b), (Zimmermann & Freksa, 1996) is based on an intrinsic reference system. Using their representation the position of a point can be described with respect to two other points. These points determine an *orientation grid* which partitions the plane into six different regions. For a point being described with respect to this grid fifteen positions can be distinguished, as shown in Figure 2.12. Such an intrinsic frame of reference is induced by two

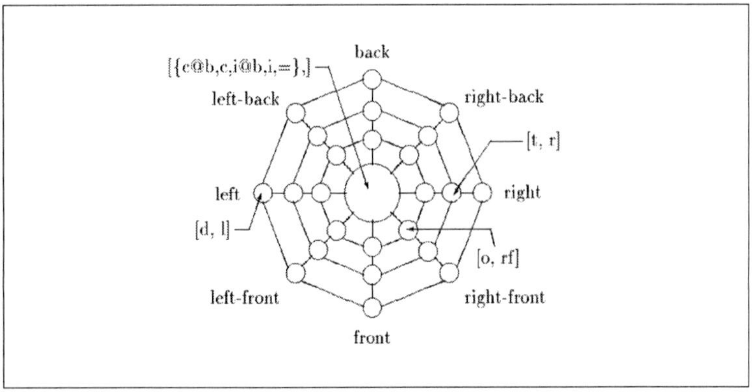

Figure 2.11: Structure integrating topological relations and orientation information — from Hernandez (1992)

of the objects present in a scene. Therefore, the objects themselves determine a context relative to which spatial relations among them can be described — no artificial external reference frame is required. However, this approach was devised in the context of navigation, and it defines reasoning mechanisms for directional information between objects, rather than describing shapes.

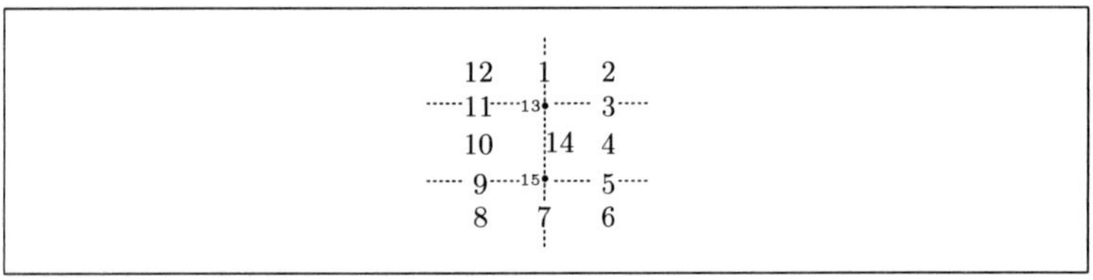

Figure 2.12: The two-dimensional orientation grid distinguishes fifteen positions

Having discussed some approaches which may aid in defining shapes, we will now consider some approaches which explicitly define shapes at the ordinal level.

Galton et al. (1999)

Similar to those sequences of components in (Egenhofer, 1997), Galton and Meathrel (1999) characterise the outlines of objects by describing the sequence of component parts. Each part belongs to a boundary type, such as *linear*, *inward pointing cusp*, or *outward pointing cusp*. A grammar for analysing any outline into its constituent components is given. In (Meathrel & Galton, 2001) they generalise their approach using a hierarchy of tokens incorporating distinct kink tokens for inward and outward pointing angles and cusps which can be used at different granularity levels. Each outline is described by a string of atomic tokens. (Figure 2.13 shows an example.) This approach is confined to describing outlines of objects: several objects forming a pattern are not considered.

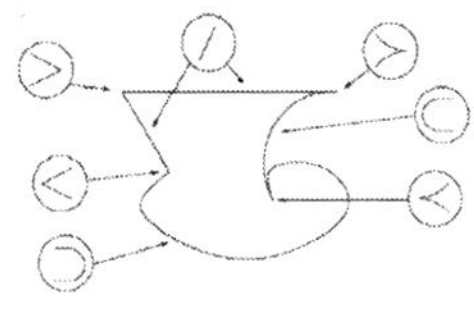

Figure 2.13: The outline of a shape and the description of its boundary parts — from Galton et al. (1999)

Jungert (1993)

An approach that is capable of determining various features of polygons, such as local extreme points, the consideration of acute, obtuse, and right angles, and the distinction between concave and convex shapes has been developed by (Jungert, 1993). In comparison to *symbolic projections* which project points only perpendicular to a given coordinate system, *symbolic slope projections* consider projections which are determined by the slopes of edges for a given polygon. By this means three vertices of two adjacent lines are projected down to the x-axis along the slope of the first line and also perpendicular to this slope, to the y-axis. The ordering of the three points with regard to the x-axis (and, separately with regard to the y-axis) is used to calculate the nature of shape features. For example, since the inside of the object is defined as being on the left when traversing the contour, a left turn means that the object is convex and a right turn that it is concave. (A straight line cannot occur, since only non-collinear points of a polygonal contour are considered). Consequently, whenever the first and second points come before the third point on the x-axis, the angle described is convex. By considering further information about the ordering of the three points in the slope projections to both the x- and y- axes, more characteristics are obtained. This shows how ordinal information (related to the ordering of points) and projections define the same level of spatial precision, as we mentioned earlier (compare Table 2.1). Likewise the previous approach, Jungert's approach considers only the description of single objects. Some examples are shown in Figure 2.14.

A generalisation of Jungert's slope projection is the *permutation sequence.*

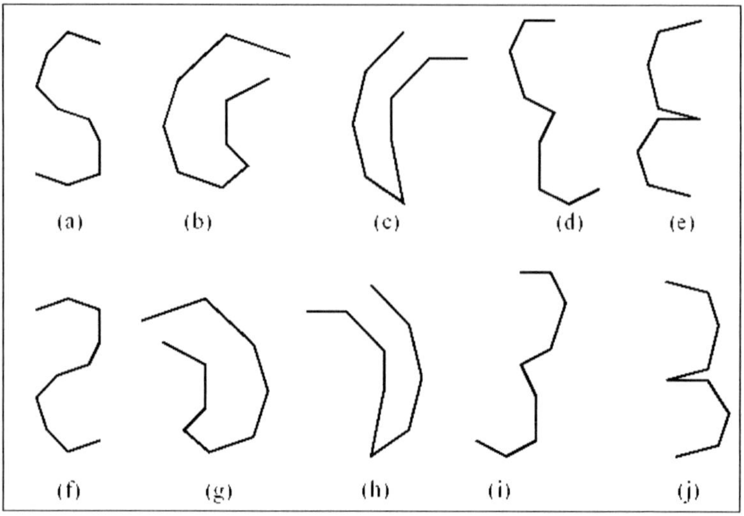

Figure 2.14: Some distinguishable polygons — from Jungert (1993)

The permutation sequence considers all different orthogonal projections of a configuration of points. Schlieder (1995a) describes the generation of the permutation sequence: and oriented line is chosen which is not orthogonal to any line connecting two of the points in the current scene (the points of a polygon, for example), and the orthogonal projection of all points onto this line determines one permutation of the point configuration. By rotating the projection-line anticlockwise the permutation will change as soon as the line passes through a slope that is orthogonal to a line which connects two of the points. All the different orthogonal projections, i.e. permutations, that will be found during a 360° rotation of the line make up the permutation sequence.

Schlieder (1996)

A qualitative concept for navigation is investigated by Schlieder (1993), which is based on the ordering of landmarks. A landmark can lie on either side of a line which is defined by two other landmarks, or it may lie directly on that line. A number of such relationships can then be used to orient oneself. For this purpose, any conceivable position can be defined by specifying the position of every landmark relative to any two other landmarks. The black area in Figure 2.15 shows an example for a region which can be addressed by the four present landmarks. As soon as one passes the line defined by two landmarks one will be in another neighbouring region. Similarly, other regions, with different shapes, can be defined by changing the landmarks' positions.

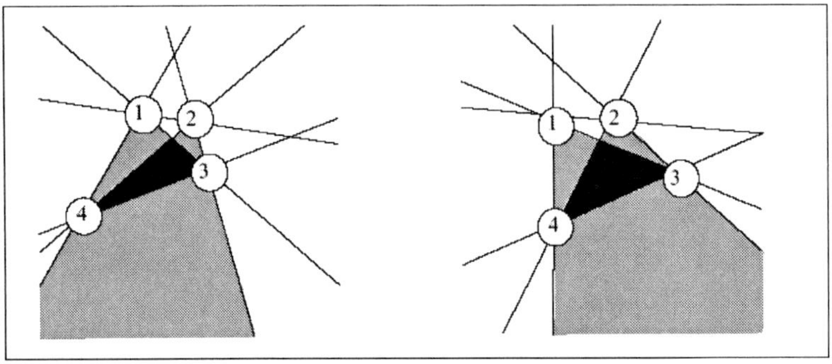

Figure 2.15: Polygonal regions described by configurations of landmarks — from Schlieder (1996)

Rather than considering the relative positions of landmarks Schlieder (1996) considers the relative positions of vertices of polygons. By this means, he derives an approach for describing polygons using ordinal information. Distinguishing the two sides of a line which is defined by two landmarks is the same as taking the triangle orientation of three vertices. A triangle is defined by three points in the oriented plane. Its orientation is defined as "+" if the path of the three points follows the mathematically positive orientation, "0" if it is rectilinear, and "-" if it follows the negative orientation. One could then consider all possible sets of three points and their triangle orientation. But when we consider shapes with n vertices, there are $\binom{n}{3}$ triangle orientations. Hence, for large n it is inefficient to proceed in this way; frequently it will be sufficient to consider only a subset of the possible point configurations. Figure 2.16 shows four similar shapes, differing only slightly in the ordering of their vertices.

This approach is of particular interest since it is possible to use it to describe the relative positions of objects, i.e. patterns of objects, and also (and equally well) the boundaries of single objects. Both single objects and patterns of objects are based on the same principle, i.e. the ordering of points in the plane. On the other hand, this description lacks means for dealing with length information. For instance, it is not possible to distinguish a rectangle from a square. Additionally, this representation is less precise than that of Jungert, who allows us to distinguish, for example, a square from a rhombus (but only considers single objects).

Linear entities

Linear entities are of particular interest to us, as we discovered when discussing characteristics of sketches. Moreover, besides positional information and orientation information, line segments implicitly encode the *length* dimension since line segments are defined not only by their position and orientation but also

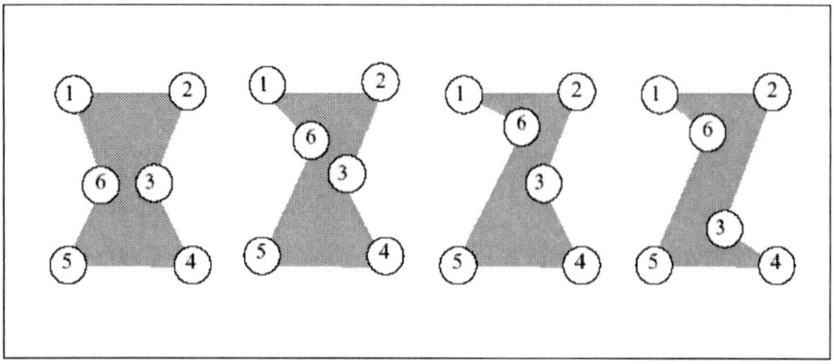

Figure 2.16: Changing ordering information between vertices changes the underlying shape — from Schlieder (1996)

by their length. Therefore, we shall have a look at those approaches which explicitly deal with linear entities at coarse spatial levels.

In the same way that Egenhofer and Herring (1991) investigated linear entities at the topological level, there have been some investigations into linear entities at the ordinal level. By contrast to Egenhofer and Herring, these approaches are all based on straight line segments, rather than on arbitrarily curved lines. Schlieder (1995b) offers an approach which describes the relative orientation of two line segments. With the aid of the triangle orientation which we have already seen applied for navigation purposes and to the description of polygonal shapes, he produces a total of fourteen line arrangements. Using a resolution of 180° this approach can be conceived as the most simple generalisation of Allen (1983) to two dimensions. Unfortunately, such a coarse resolution does not allow for some distinctions which are obviously perceptually aided, as shown in Figure 2.17.

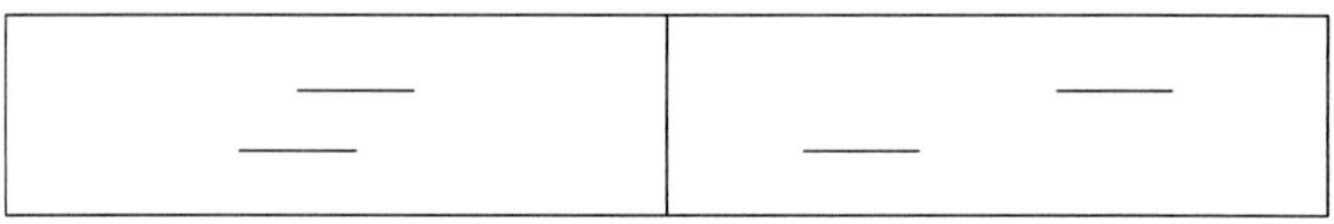

Figure 2.17: These two bipartite line arrangements are not distinguishable using the approach of Schlieder (1995b)

Moratz, Renz and Wolter (2000) specify 24 relations, which extend the relation-set of Schlieder (1995b) by also considering the equality of endpoints. Further relations are obtained by an extension where one endpoint of a line segment can be precisely behind the other line segment, on that line segment, or in front of it, leading to a final total of 69 different relations. Unfortunately, these still do not distinguish some line arrangements, such as those in Figure 2.18. In this example, it holds for all three cases that the start point and the end point of B are on the right of A, and that the start point and the endpoint of A are both on the left with respect to B. Using their notation, we obtain A *rrll* B for all three cases. Only for the special case where A lies directly on the line which is defined by B are the corresponding arrangements distinguishable.

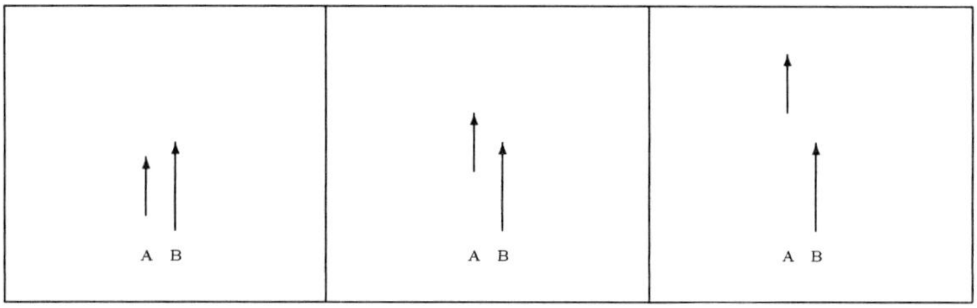

Figure 2.18: These three bipartite line arrangements are not distinguishable by the dipole approach of Moratz et al. (2000) — each arrangement is characterised by A *rrll* B

Renz (2001) confines himself to 26 relations which are equal to the 13 relations of (Allen, 1983), but he also considers their converse directions.

2.2.3 Summarising spatial representations

When discussing the importance of shapes in sketches we have learned that we have (for linear entities) to take into account position, orientation, and size simultaneously, as well as the relationships between them. We have examined existing qualitative spatial representations against these requirements and found, as the summaries in the following tables show, that there is currently no approach which satisfies all our requirements.

When discussing different approaches we learned that topological relations are perceptually aided, but also that topological distinctions are too coarse. Having identified the level of affine transformations, on the other hand, as being too fine, only approaches at the level of ordinal information (projections) remain (see Table 2.2).

Most qualitative representations, in particular those which are based on points or regions (see Table 2.3) are capable of providing some means for describing patterns. By contrast, approaches which are based on linear entities are confined to describing the shape of single objects ((Cohn, 1995), (Galton & Meathrel, 1999), (Jungert, 1993)), and they all lack means for considering size or respectively length information. Size and distance information is only treated in theories which do not deal simultaneously with shapes, for example

Table 2.2: QSR approaches and their spatial precision

Approach	Topological	Ordinal
Guesgen 1989		x
Egenhofer, Franzosa 1991	x	
Egenhofer, Herring 1991	x	
Randell et al. 1992	x	
Freksa, Zimmermann 1992		x
Hernandez 1992	x	x
Jungert 1993		x
Cohn 1995	x	
Schlieder 1995, 1996		x
Egenhofer 1997	x	x
Galton, Meathrel 1999		x
Moratz et al. 2000		x
Renz 2001		x

Table 2.3: QSR approaches and their basic entities

Approach	Points	Lines	Regions	Others
Guesgen 1989	x			respective rectangles
Egenhofer, Franzosa 1991			x	
Egenhofer, Herring 1991	x	x	x	
Randell et al. 1992			x	
Freksa, Zimmermann 1992	x			
Hernandez 1992	x		x	
Jungert 1993		x		
Cohn 1995			x	
Schlieder 1995, 1996	x	x		
Egenhofer 1997		x	x	
Galton, Meathrel 1999		x		differently curved primitives
Moratz et al. 2000		x		
Renz 2001		x		

that of (Hernandez, Clementini & Di Felice, 1995). Distance information is especially important for dealing with patterns and objects with complex shapes. (Schlieder, 1996) also lacks means for dealing with size and distance information, but in contrast to the others his approach can readily be extended to deal with patterns.

If we exclude approaches which function at the topological level, and those which are not based on lines, what remains are those approaches which are capable of dealing with linear entities in shapes, namely (Schlieder, 1996), (Jungert, 1993), and (Galton & Meathrel, 1999); and those which deal with linear entities in general, namely (Schlieder, 1995b), (Moratz et al., 2000), and (Renz, 2001). However, as demonstrated in section 2.2.2 (see Figs. 2.17 and 2.18) these approaches do not consider all distinctions which are obviously perceptually aided. Confusingly, (Moratz et al., 2000) and (Renz, 2001) do consider some distinctions which are *not* perceptually aided, since they distinguish positions and orientations which are precisely aligned. Above all, the complex relationships between the position, orientation, and size dimensions have not so far been addressed for handling imprecise shape information (and linear entities in particular) in pictorial space.

2.3 Object of research

We have seen that sketches are imprecise, and that there are several reasons for this. Before a sketch is made, one usually remembers or imagines the shape of an object — but spatial information about shapes which have been recalled from memory is frequently distorted, so are perceptions in general, and, of course, peoples' sketching skills are limited. Current sketching systems either avoid comparing shapes completely, or else they are confined to small and simple object domains. A broader area of theories for dealing with imprecision concerning spatial information is provided in the field of qualitative spatial reasoning. It is therefore worthwhile to investigate how well theories in this field cope with imprecision in sketches.

Having done this in the last section, we conclude that existing qualitative approaches lack means for representing position, orientation, and size (the fundamental dimensions for coping with shapes in the plane) simultaneously. In order to characterise sketches we have to describe shapes. By contrast, the two main application areas for qualitative spatial reasoning approaches do not require shape information. In navigation tasks information about direction (orientation) from one point to another is of primary interest, and in geographical information systems either orientations between point like abstractions of objects or adjacency relationships between regions are important. This is the reason why there is not yet any qualitative representation which is capable of dealing with shapes in a comprehensive way. The question arises whether it may prove necessary to fall back upon metrical descriptions since neither topological relations nor ordinal relations are currently capable of extensively describing imprecise shapes. To take up this challenge we will start by considering linear entities, which are the simplest objects capable of showing variations in all of the fundamental dimensions of pictorial space, namely position, orientation, and length.

We earlier identified linear entities as being an important (probably the most important) ingredient in sketches. The significance of linear entities becomes especially clear when considering any sketch (in particular of complex, non-linear objects) made with a pencil. For instance, a region with a complex shape will be made up of a linear curve. It seems logical to base a sketch representation on linear entities, especially when using qualitative representations in order to cope with imprecision, i.e. with the distinction between necessary and accidental properties, and we shall therefore model the necessary distinctions of sketches using qualitative arrangements of lines. This will require us to solve or avoid the shortcomings of other approaches which are also based on lines. Especially, we will need to find a way of making all those distinctions which are perceptually aided. For this purpose, we will investigate how lines can be arranged in the plane while considering perceptually aided distinctions similar to those made by Allen (1983).

Part II

Theory

Chapter 3

Qualitative line arrangements

In this chapter we will investigate line arrangements from a qualitative point of view. The proposed representation and reasoning processes rely on those sort of distinctions which are easily comprehensible by natural perceptual systems. For instance, it will be possible to distinguish whether two lines are on the same side with respect to another one, or whether they are on different sides — but we will not be able to obtain their precise positions. From an information theory point of view, possible arrangements of lines are systematically investigated, making use of distinctions which are both easily perceivable and easily sketchable: *left* and *right*, *back* and *front*, *during* and *contains*. We shall learn that such coarse classifications will nevertheless allow a vast number of distinctions to be made, and finally arrive at a theory of line arrangements which is based on a small set of relations between intervals within the plane. This theory forms the basis for a *qualitative geometry* using which we will consider the qualitative characterisation of polygons in the next chapter.

Intervals form the basis for the proposed qualitative representation. Depending on the context, they represent either distances between objects or approximations to a part of a shape, which we then refer to as lines. In the following, the word lines is used to mean straight line segments rather than arbitrarily curved lines. Lines are the most simple entities which encode both orientation and length. Additionally, as soon as a line is placed somewhere in the plane it occupies a location. We then find all fundamental dimensions which are needed in order to describe shape information united in one single entity: each line encodes information about position, orientation, and length, lines being the simplest entities which encode information about all these dimensions. Indeed, without using external reference systems we need at least two lines in order to determine position, orientation, and length of one line (relative to the other one). In what follows, lines (or, more generally, intervals) will be shown to be particularly useful entities, providing a basis for the description of pictorial information.

Any shape in the picture plane can be described in terms of the arrangement of a number of lines. This holds in particular for sketches, in as much as their rough structures can be adequately described by approximating them with a number of lines. Hence, any sketch may be described by an arrangement of lines, and we will introduce a representation that readily allows us to characterise perceptual distinctions, providing a means for distinguishing between necessary and accidental shape properties, these distinctions being based on what we refer to as *qualitative line arrangements.*

3.1 Bipartite arrangements

The simplest arrangements are made up of two lines. In this section we devote ourselves to these *bipartite arrangements*, and investigate how two lines can be related according to (Gottfried, 2004a). For this purpose, we take a look at the Allen relations (shown in the Figure 3.1) once more.

before	$<$
meets	m
overlaps	o
starts	s
finishes	f
during	d
equals	$=$
contains	di
is finished by	fi
is started by	si
is overlapped by	oi
is met by	mi
after	$>$

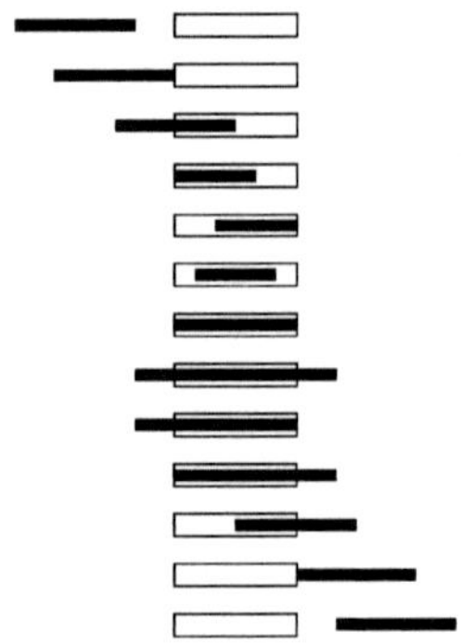

Figure 3.1: Allen's thirteen one-dimensional interval relations

These thirteen relations can be transformed into each other, either by changing the relative position of the intervals, for example, in order to transform *before* into *meets*, or by changing their relative length, for example, in order to transform *during* into *equals*. However, we are dealing with a two-dimensional image plane. Objects within the plane have been thoroughly investigated on the basis of Euclidean geometry, and we are accustomed to how things are described by Euclidean geometry. Considering the possible ways in which one could arrange intervals in the two-dimensional plane, one conclusion is self-evident. In contrast to the one-dimensional case where it is only possible to displace intervals either forwards or backwards, in two dimensions it is additionally possible to displace intervals in the orthogonal direction; that is, sideways. This is illustrated in

Figure 3.2.b. Each arbitrary displacement can then be constructed by a pair of two displacements, one horizontal and one vertical displacement. Figure 3.2.c shows horizontal and 3.2.b vertical displacements.

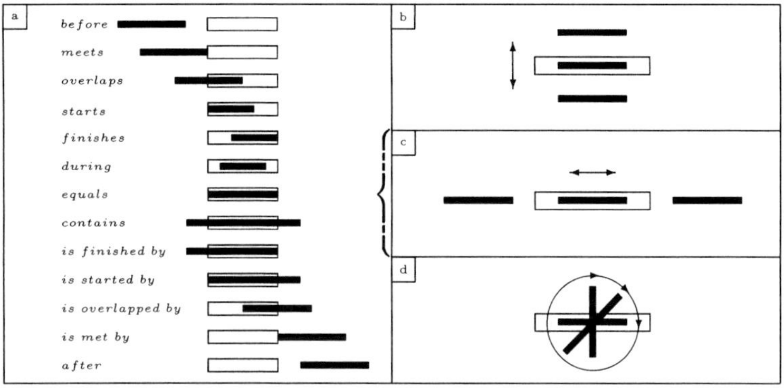

Figure 3.2: Left: Allen's thirteen one-dimensional interval relations; Right: Further variations are possible in the two-dimensional plane

Figure 3.2.d shows yet another variation possible in the two-dimensional plane. In the one-dimensional case it is not possible to rotate an interval (for instance, around its midpoint or around an endpoint). But in two dimensions this leads to additional variations connected to the orientation of intervals within a range of 360°. Finally, the length of intervals can be changed, as is already possible in one dimension. By means of these types of transformation, every conceivable arrangement of two intervals can be constructed, although we aim to distinguish only those arrangements which are as easily distinguishable as the Allen relations. For example, we want to distinguish between the two arrangements on the right-hand side of Figure 3.3, but not between the pair on the left.

The idea is that we will ultimately obtain a number of relations which are easily distinguishable by perception. There exist infinitely many different arrangements of two intervals from a metrical point of view. Are there as many from the perceptual point of view? Considering the one-dimensional case and the example in Figure 3.3 it seems likely that there are many fewer qualitative arrangements.

3.1.1 Reference system

In order to systematically analyse what kinds of arrangements are possible, it is necessary to define a reference system that can differentiate between all the

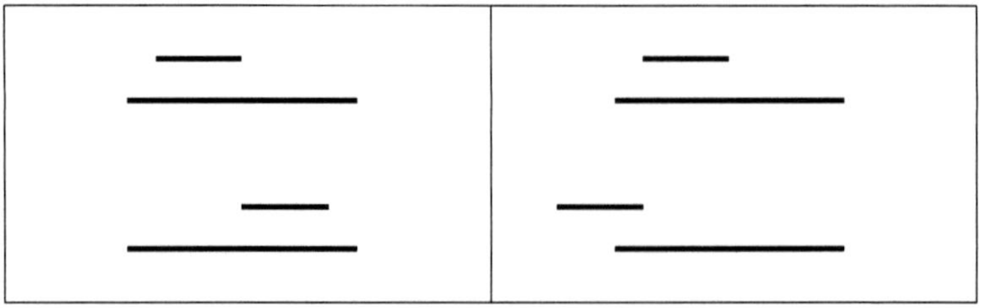

Figure 3.3: Two pairs of more or less equal bipartite arrangements

arrangements which we wish to distinguish. Taking the 13 Allen relations as an example, such a reference system has to distinguish precisely those positions which define each relation. As one interval can be regarded as a *reference interval* relative to which the other interval is described, the reference system we are looking for is defined in terms of an interval. The endpoints of this interval define that reference system, as depicted in Figure 3.4.

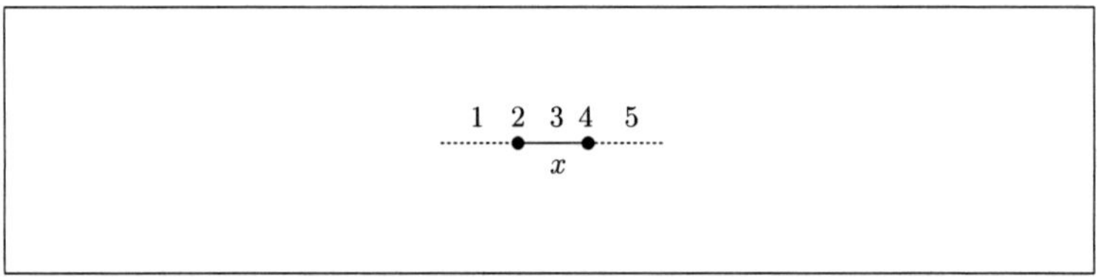

Figure 3.4: The one-dimensional reference system distinguishes five positions

The reference interval x in Figure 3.4 lies from point 2 to point 4. In one dimension it determines altogether 5 different locations relative to which another interval, the *primary interval*, can be described. Placing the endpoints of the primary interval at every possible combination of these locations and discounting the two possibilities that produce a point rather than a line, we obtain the 13 Allen relations.

As the orthogonal direction allows for those vertical displacements in Figure 3.2.b, an orthogonal dimension is introduced at locations which are again determined by the endpoints of the reference interval. In this way, we obtain the two-dimensional reference system in Figure 3.5. This reference system equals that introduced by Christian Freksa and Kai Zimmermann to which they refer to as the *orientation grid* ((Freksa & Zimmermann, 1992), (Freksa, 1992b), (Zimmermann & Freksa, 1996)). While we are interested in describing the relative position of two intervals, they use this system in order to describe the position of a point in relation to two other points, the latter two defining the reference system. Fifteen positions can then be distinguished for the primary point. As we shall see below, there are many more possibilities for describing the position of the primary interval when using this reference system, since an interval is defined by two points and the positions of both of these must be considered simultaneously.

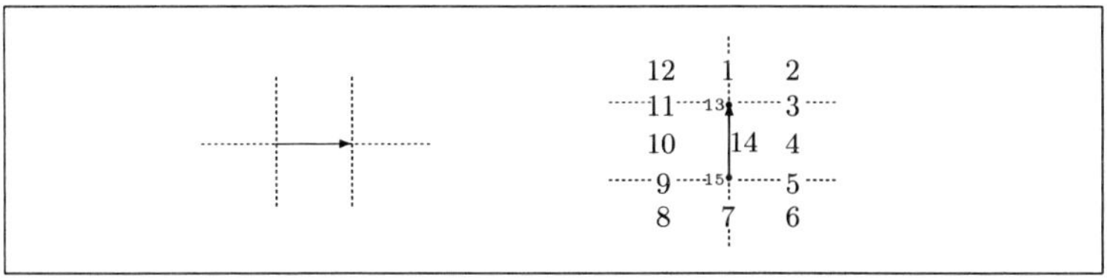

Figure 3.5: The two-dimensional reference system distinguishes fifteen positions; the reference interval has been drawn continuously — on the right hand side it has been rotated by 90° so that the left and right of the reference system corresponds to the left and right in the picture plane

Using the orientation grid we are able to systematically consider all conceivable relations between intervals in two dimensions. For this purpose, we place both endpoints of the primary interval successively at each of those 15 positions depicted in Figure 3.5, yielding $15^2 = 225$ arrangements which are shown in Figure 3.6. Each such bipartite arrangement defines a relation between two intervals, and we can assign any pair of metrically determined intervals uniquely to one such relation. We refer to these relations as to *bipartite arrangements*, and denote the set of these relations by $\mathcal{BA}$. Note that the position of the primary interval is completely determined with respect to the reference interval without requiring any external reference system. This is in contrast to the approach of (Schlieder, 1995b) who defines line segment relations on the basis of an oriented plane. Using this approach, it is possible to decide for three distinguishable points whether they are oriented clockwise or anticlockwise regarding the plane, and two line segments are characterised by four such triangle orientations which describe the configuration of the endpoints of the two line segments. The relations of $\mathcal{BA}$ do not require reference to any external system but only that the two endpoints of the reference interval be distinguished.

In Figure 3.6 the reference interval is always the vertical line segment which is oriented as shown by the first arrangement at the top-left. Its endpoints define different locations around it, as shown on the right hand side of Figure 3.5. The fifteen relations in one row in Figure 3.6 can be described as follows. Within each row, the position of one endpoint of the primary interval is kept fixed, while the other endpoint is placed in each position in turn, in the order defined in Figure 3.5. When the moving endpoint has visited all 15 positions, the fixed endpoint is moved one place clockwise for the start of the next row, and further rows are constructed in the same way. Positions 15 and 13 in Figure 3.5 correspond to the endpoints of the reference interval. Positions denoted by 1, 3, 5, 7, 9, 11, and 14 lie on the orientation grid, and the remaining positions (2, 4, 6, 8, 10, and 12) correspond to six different regions. A point lying in one of these regions is said to be in general position, and all others are in singular position. Intervals in singular position are depicted by bulky points in Figure 3.6. We denote the position of the primary interval y with respect to the reference interval x by x_y, and it holds that

$$x_y \in \mathcal{BA} = \{\mathcal{BA}(i) | i = 1, 2, ..., 225\} \tag{3.1}$$

Table 3.1: Allen's relations are a subset of $\mathcal{BA}$; further symmetrical relations of $\mathcal{BA}$ are not considered (for instance, $217 \equiv 105$)

Allen	$\mathcal{BA}$
before	1
contains	7
meets	13
overlaps	14
is finished by	15
after	97
is overlapped by	104
is met by	105
is started by	187
starts	194
equals	195
during	209
finishes	210

A prominent subset of $\mathcal{BA}$ equals Allen's time intervals. Actually, projecting each of the 225 relations of $\mathcal{BA}$ orthogonally onto that line on which the reference interval lies, each relation equals one of Allen's time intervals. Only the positions of the related line's endpoints perpendicular to the reference line distinguishes the two-dimensional from the one-dimensional case. These endpoints can be placed on either side of the reference line. As is shown in Table 3.1, for the thirteen Allen relations, $\mathcal{A}$, it holds that $\mathcal{A} \subset \mathcal{BA}$.

Two related approaches define equivalence classes among the relations of $\mathcal{BA}$. Schlieder (1995b) obtains 14 line segment relations in general position and 63 when singular positions are included. The dipole relations of Moratz et al. (2000) define a set of 24 and an extended set of 69 relations. As the relations of both Schlieder and Moratz form equivalence classes containing several different $\mathcal{BA}$-relations they describe coarser grained sets of relations.

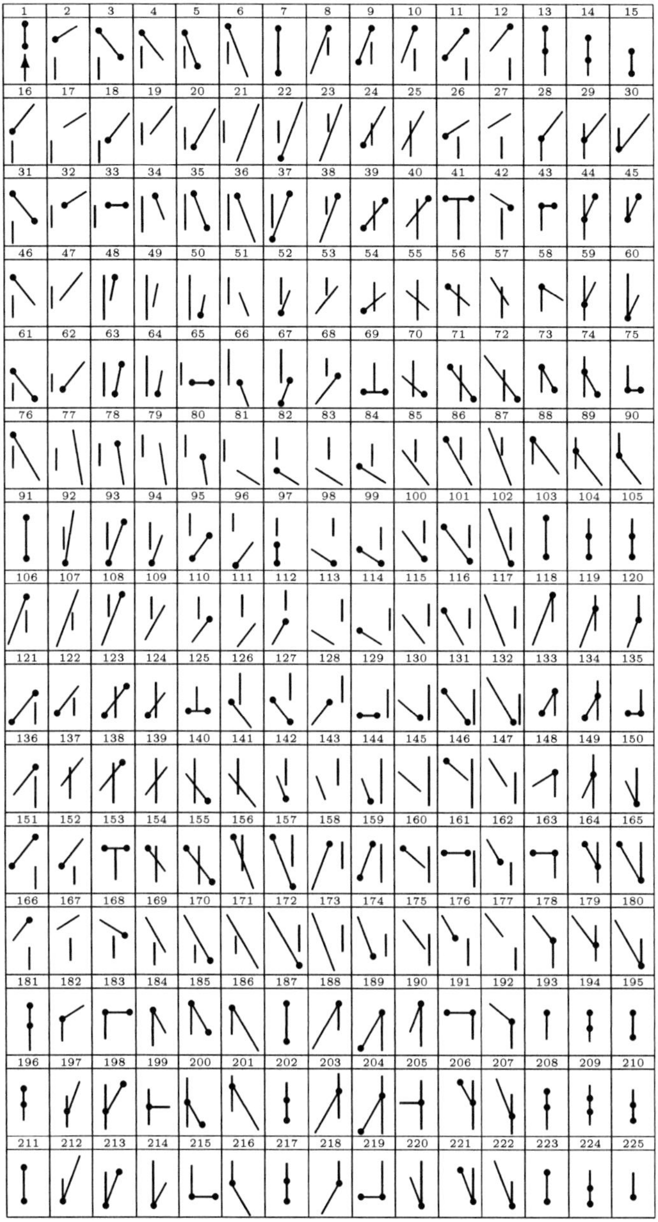

Figure 3.6: There are 225 possible two-dimensional bipartite arrangements if the two endpoints of the two intervals are distinguishable; the bulky points denote singular positions of the endpoints of the primary interval which lie on the reference system

Regardless of whether all possible $\mathcal{BA}$ distinctions are made or only a subset, the meaning of these relations are as follows. For any image, an interval x is defined by a pair of points in $\mathbb{R}^2$. By *interval interpretations* we mean pairs of distinct real points in $\mathbb{R}^2$ which metrically determine intervals, i.e. the underlying domain of intervals is infinite. Given two interpreted intervals, x and y, exactly one relation of $\mathcal{BA}$ holds for x_y, and exactly one relation holds for y_x. Conversely, a formula $x_y = \mathcal{BA}(i), i \in \{1, 2, ..., 225\}$ with the intervals x and y is said to be satisfied by an interpretation if that interpretation realises the relation $\mathcal{BA}(i)$ in $\mathbb{R}^2$. The relations of $\mathcal{BA}$ subsume infinitely many interpretations, and each relation of $\mathcal{BA}$ can be regarded as uninterpreted. We call such uninterpreted relations *qualitative*, and any set of relations of $\mathcal{BA}$ is called a *qualitative line arrangement.* We are now able to state our Thesis more precisely:

> **Thesis:** *Perceptually aided and mentally aided distinctions are represented by qualitative line arrangements. Metrically aided distinctions include different interpretations of one and the same qualitative arrangement.*

3.1.2 Position versus orientation

So far we have only taken into account positional relations. In addition to these variations, the orientation of the primary interval can be changed, as illustrated in Figure 3.2.d. In fact, orientation and positional information are interrelated. Some relations which are defined solely on specific positions constrain possible variations in orientation. $\mathcal{BA}(1)$, for example, allows only two orientation variations which differ by exactly 180°, while Figure 3.7 shows that $\mathcal{BA}(171)$ allows only orientation variations within a range of less than 90°, whereas $\mathcal{BA}(17)$ can be oriented arbitrarily, i.e. every orientation variation within a range from 0° to 360° is possible. This demonstrates that positional information and orientation information are mutually dependent. This dependence can be stated more generally by clarifying how position and orientation as well as location and direction are interrelated:

$$\frac{position}{orientation} = \frac{location}{direction} \tag{3.2}$$

An object has a position, i.e. it occupies a location, and it has an orientation, i.e. it points into some direction. An oriented reference interval defines different locations, as shown in Figure 3.8.a. The position of another interval, the primary interval, can be described with respect to these locations. Similarly, an oriented reference interval defines different directions, as shown in Figure 3.8.b. The orientation of the primary interval can be described with respect to these directions. In the same way that the reference interval uses 90° angles for partitioning the plane into sections that define different locations, different directions are defined by partitioning the range of possible directions into 90° angles, as shown in Figure 3.8.b. In this Figure, eight distinguishable directions

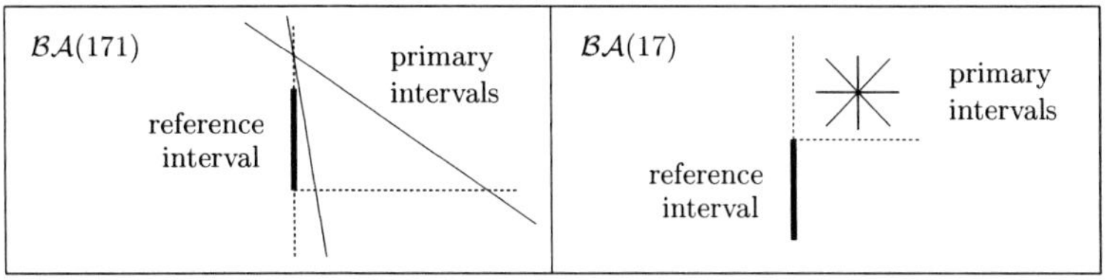

Figure 3.7: Left: the orientation of the primary interval can be changed only within a range of less than 90°; Right: the primary interval can be oriented arbitrarily

are denoted by the corresponding locations of Figure 3.8.a, showing the relationship between location (position) and direction (orientation). The reference interval simultaneously determines possible locations and directions , so position and orientation of the primary interval cannot be treated independently: the orientation of the primary interval is related to its position, as each direction is related to specific locations which are defined by the reference interval.

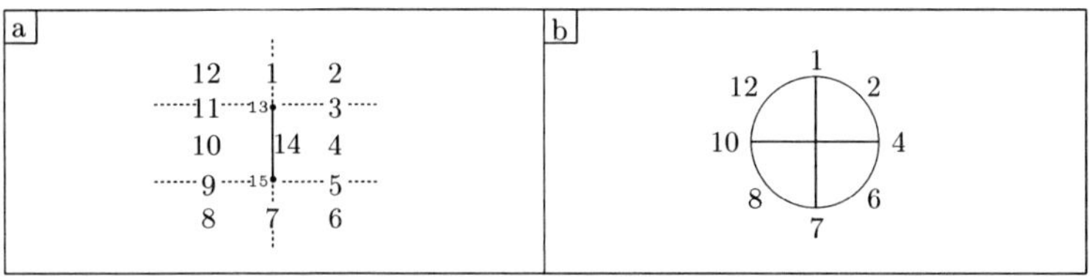

Figure 3.8: (a) The two-dimensional reference system distinguishes fifteen positions — the reference interval runs from point 13 to point 15; (b) as for the locations different orientations are defined taking a resolution of 90° angles

Finally, there is the length of the intervals, which cannot be treated independently of positional information either. For example, solely by changing the length of the primary interval, $\mathcal{BA}(17)$ can be transformed into $\mathcal{BA}(27)$ and vice versa. $\mathcal{BA}(1)$ can be arbitrarily lengthened in one direction without changing the relation. By contrast, if it is extended in the other direction we will eventually obtain $\mathcal{BA}(13)$. Thus, all three dimensions (length, direction, and location) are mutually dependent.

Considering variations in length, direction, and location of intervals, we can readily distinguish a large number of different bipartite relations, and it is necessary to ask ourselves whether all these distinctions are actually useful. From the point of view of sketching in particular, a smaller number of distinctions is probably sufficient, as some relations (such as $\mathcal{BA}(2)$ and $\mathcal{BA}(17)$) are difficult to distinguish in a sketch. Combining the positional relations of $\mathcal{BA}$ with the orientation variations we obtain quite a large set of relations. There are 225 possible positional relations and 8 orientation variations, making a total of 225*8=1800 relations. Each such relation encodes simultaneously information about position and orientation. Since many relations, such as $\mathcal{BA}(1)$ and $\mathcal{BA}(2)$, allow only a subset of the orientation variations, there are ultimately more than 225 but fewer than 1800 relations. Comparing them with both Schlieder (1995) and Moratz (2000) it can be seen that relations exist in $\mathcal{BA}$ that cannot be distinguished by either Schlieder or Moratz. The arrangements in Figure 3.9, for example, although these arrangements are certainly simple to distinguish both by perception and also when sketching them. On the other hand, we are always able to distinguish by $\mathcal{BA}$ the relations identified by Schlieder[1] and Moratz[2]. Analysing the relations of $\mathcal{BA}$ more thoroughly in the next sections, we will produce a clear and rather handy set of relations.

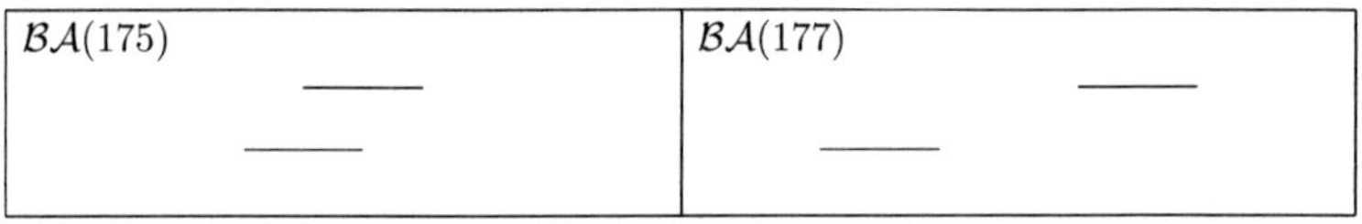

Figure 3.9: These two bipartite line arrangements (both primary intervals pointing to direction 1 (see Figure 3.8)) can be distinguished by $\mathcal{BA}$, but not by either Schlieder (1995) or Moratz et al. (2000)

3.1.3 Omitting singularities and intersections

It is important to note that all these relations can be divided into two general classes; those which involve singular positions and those which do not. Singular positions concern precisely aligned intervals, and are therefore somewhat inappropriate when dealing qualitatively with objects, in particular when trying to distinguish them in a sketch. Qualitative distinctions concern relations about which we feel confident. Such relations represent coarse knowledge, implying that small changes are negligible. But arbitrarily small changes to intervals in singular position violate this property of qualitative relations. A relation in singular position must not be changed even the smallest amount, or the relation

[1] Schlieder's line segment relations in general position are represented by $\{\mathcal{BA}(i)|i \in \{21^7, 49^4, 49^{10}, 21^1, 111^4, 167^4, 55^4, 55^{10}, 167^{10}, 111^{10}, 117^1, 145^4, 145^{10}, 117^7\}\}$ - the top-index indicating the orientation accordingly to Figure 3.8.

[2] $\mathcal{D}_{24} = \{\mathcal{BA}(i)|i \in \{35, 49^4, 49^{10}, 63, 83, 55, 27, 167, 139, 111, 131, 145^4, 145^{10}, 159, 191, 183, 163, 43, 219, 215, 135, 75, 223, 195\}\}$

$\mathcal{D}_{69} = \mathcal{D}_{24} \cup \{\mathcal{BA}(i)|i \in \{161^{10}, 113^8, 113^2, 161^4, 149, 166, 125, 153, 98, 199, 2, 52, 142, 12, 94, 16, 41, 48, 205, 59, 81^{12}, 69, 81^2, 17, 1^1, 181, 196, 91, 211, 103, 97, 105, 104, 209^7, 224, 208, 1^7, 13, 14, 7, 15, 187, 209^1, 194, 202\}\}$

will no longer hold. In order to consider such positions in drawings, tools such as rulers and set squares are needed, but rough sketches are made without such tools. For this reason we must leave out relations including singular positions. However, we do not simply exclude those relations but instead reassign them to other relations. For a relation such as $\mathcal{BA}(2)$, this can be achieved by excluding endpoints of the primary interval which lie in singular position (making $\mathcal{BA}(2)$ into $\mathcal{BA}(17)$). Relations which are entirely in singular position (like $\mathcal{BA}(1)$) can be reassigned by placing the endpoint of the primary interval in neighbouring regions which enclose that singular position (in this case, $\mathcal{BA}(27)$). Another general and convenient way of dealing with singularities is discussed later.

There is a second issue to be addressed. We must not confuse the information contained in an image with the interpretation of that image. In particular, this is important when considering the overlap of objects. $\mathcal{BA}(55)$ provides an example. Such an arrangement is normally interpreted as representing an intersection of two intervals. But from the point of view of an image, at any location there can only be information about one colour. Intersections or overlaps are not realisable in the two-dimensional plane. Whatever we regard as intersections is essentially the three-dimensional interpretation of what is actually flat pictorial information. From this it follows that intersections like $\mathcal{BA}(55)$ can be excluded at the level of basic pictorial relations. At more abstract levels, however, we may be interested in representing intersections. For this purpose, we shall later show how intersections can be represented by means of arrangements of intersection-free relations.

Neither Schlieder nor Moratz make a distinction between singular relations and general relations but treat them at the same representational level. The same holds for intersections. We discuss in the next section what kinds of arrangements remain when confining the set of basic relations to those which are both in general position and free of intersections.

3.2 Disconnected arrangements

We now have a number of qualitative line arrangements describing different forms of disconnection. Choosing two lines with a desired length, and using a resolution which distinguishes 90° angles within the Euclidean plane, the lines can be arranged in 125 different ways. The set of these relations is denoted by $\mathcal{BA}_{23}^{8}$, indicating 23 different positional relations and 8 distinguishable orientation relations.

Figure 3.11 shows these relations arranged in their neighbourhood graph. Both the reference interval and the primary interval are oriented, i.e. their endpoints are distinguishable (see relation F_l in Figure 3.11). The endpoints of the reference interval are as depicted in Figure 3.8.a, with the eight possible orientations of the primary interval arranged around each relation in Figure 3.11 as shown in 3.8.b. Two positional relations are connected if they can be transformed into each other, by changing the position of one or both endpoints of the primary interval without passing through any third relation. By contrast to

$\mathcal{BA}_{125}$, Figure 3.12 shows the graph of $\mathcal{BA}_{113}$, which omits singular positions; whereas $\mathcal{BA}_{113}$ can be appropriate for describing perceptual processes which cannot unerringly detect singularities, the singularities in $\mathcal{BA}_{125}$ (such as F_m^F) can be useful for designing configurations, in which singular relations may be useful. Further neighbourhood graphs are conceivable by taking into account the orientation of the primary interval as well as its position. A minimised version of the neighbourhood graph is depicted in Figure 3.10, distinguishing only the different positional relations.

A mnemonic description, shown on the right hand side in Figure 3.10, allows us to concisely indicate the various relations. These mnemonics aid in comprehending their meaning. The identity relation is in the middle of the neighbourhood graph. For two intervals the identity relation holds if the two intervals are equal. Since each interval is uniquely defined in the picture plane and no overlap between intervals is possible, no transformation is possible between the identity relation and any other relation. (Thus, there is no connection to the identity relation in the neighbourhood graph.)

The relationship between position and orientation is again shown by the way in which positional relations constrain possible orientation. Twelve of the 23 positional relations allow all orientation variations: two relations, C_l and C_r, allow six different orientations, eight relations (FO_{ml}, FO_{mr}, FC_l, FC_r, BC_l, BC_r, BO_{ml}, and BO_{mr}) allow only two orientations each, and the identity relation allows exactly one orientation. The combination of the resolution (90° angles and six regions) and these dependencies between position and orientation result in a final total of 125 distinguishable relations.

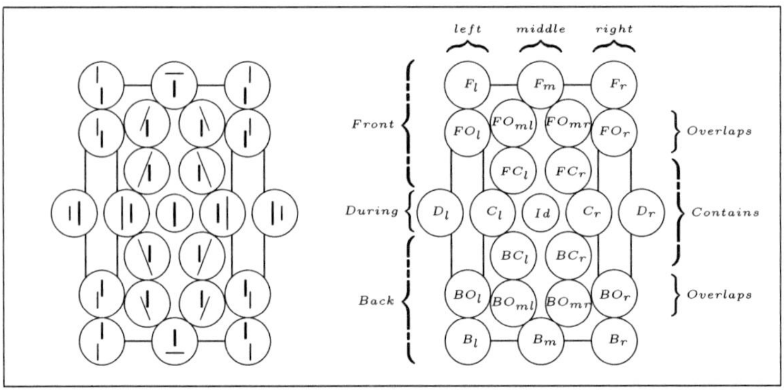

Figure 3.10: Left: Interval relations embedded in two dimensions; the vertical reference interval is displayed bold. Right: The mnemonic depiction of the interval relations

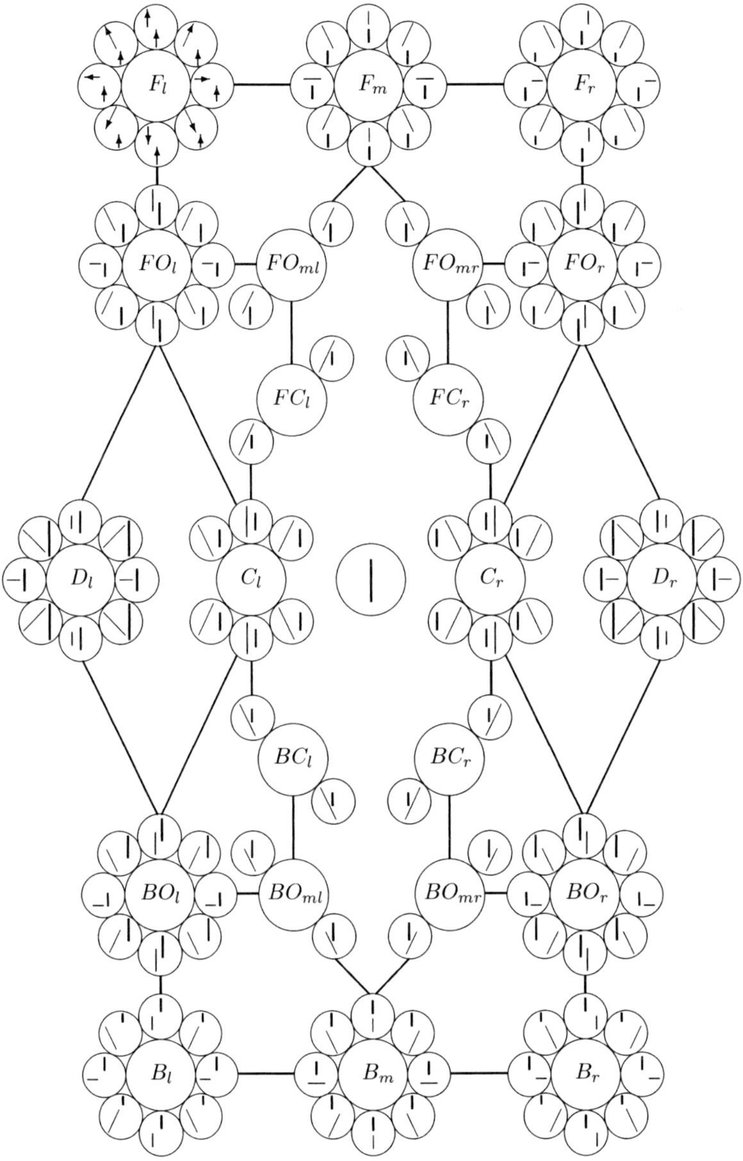

Figure 3.11: The 125 relations of $\mathcal{BA}_{23}^{8}$

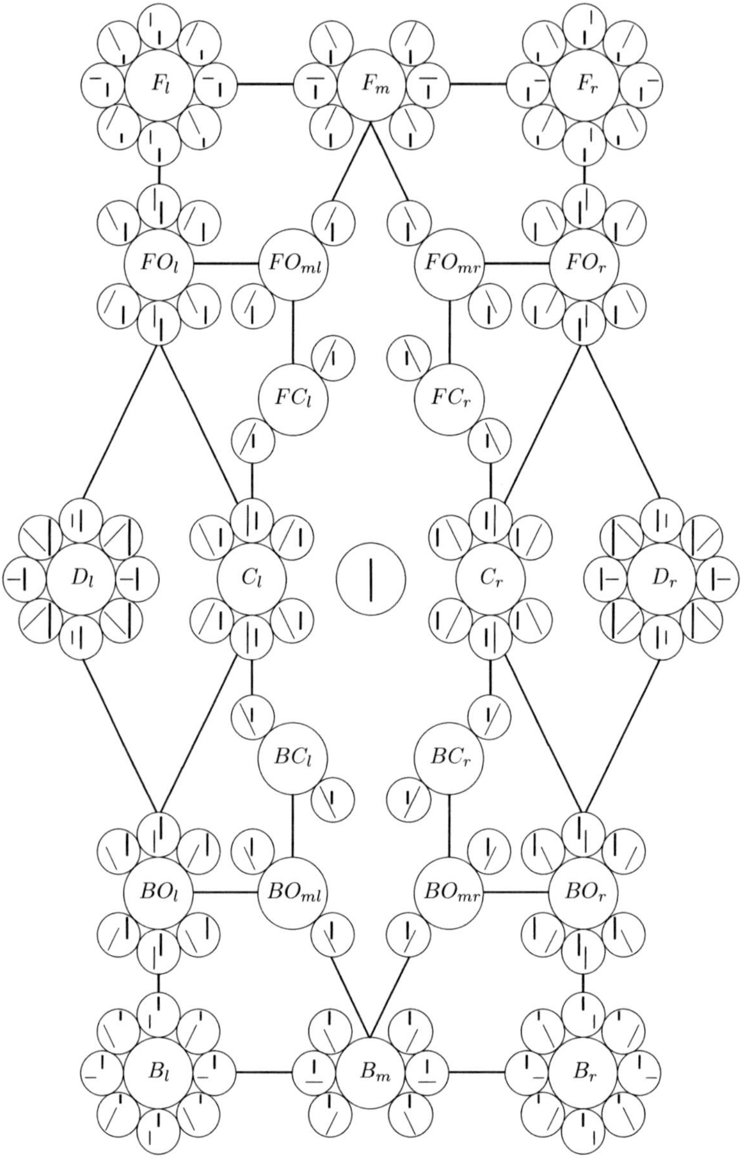

Figure 3.12: The 113 general relations of $\mathcal{BA}_{23}^{8}$

Any reasoning procedure must deal in some way with subsets of $\mathcal{BA}_{23}^{8}$. When we know exactly which relation holds between objects we are concerned with a subset containing only one element. But after a single inference step our knowledge frequently becomes indeterminate, and this indeterminacy means that we are concerned with subsets containing more than one relation. Where this is the case, we must take into account all those relations which may hold. Sometimes our knowledge is completely indeterminate, and every relation in $\mathcal{BA}_{23}^{8}$ could hold. We can exclude relations if there are reasons why these relations cannot hold in the given situation. Subsets of $\mathcal{BA}_{23}^{8}$ can be illustrated concisely in iconic form; Figure 3.13 shows some examples — each icon is a minimised depiction of the graph in Figure 3.10. Positional relations contained in a subset are printed in black, and orientation relations can be specified for each positional relation separately. The identity relation can often be omitted from the iconic depictions, increasing the clarity of the diagram.

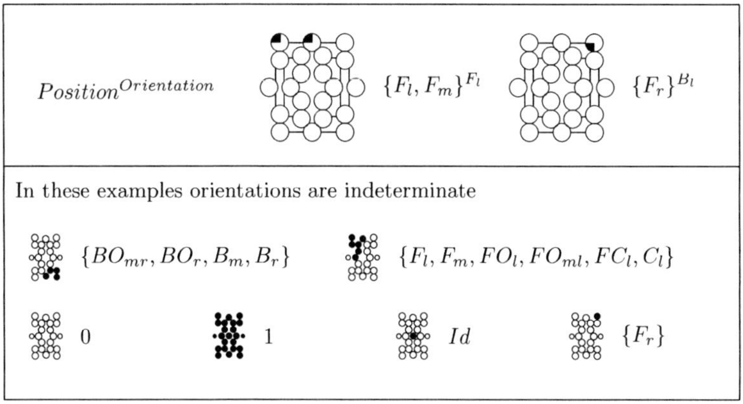

Figure 3.13: Examples for the iconic depiction of sets of relations

3.2.1 N-partite arrangements

Nothing can be said about the position of a single line segment without external reference. Once there is a second line to fulfil the role of the reference line, the former line's position can be determined, though only accordingly to that reference line. But the more lines there are, the more accurately the position of a line can be described, as a rule. The arrangement of lines also affects how precisely the position of a line can be described. If, for instance, an arbitrarily large number of lines are arranged in such a way that the primary line is in the same relation to all reference lines, the position of the primary line cannot be determined very accurately. By contrast, the more different relations there are between the reference lines and the primary line, the more accurately the position of the primary line can be determined. Consider, for example, a regular partitioning of the plane into a number of small squares, all the lines which form these squares having the same length d. The shorter d is, the more precisely the position of the primary line can be described. The regular partitioning of the plane actually forms a metric with the minimal distance d. Thus reference segments introduce a special context, and the position of a primary line can be described with regard to this context, with each primary line being part of the context for any other line segment. Thus, an arrangement of lines is described in a self-referring way.

Any conceivable arrangement of n lines can be described by considering all $\binom{n}{2}$ pairs of lines. To be more precise, there are $\frac{n!}{(n-2)!}$ bipartite arrangements since a relation of $\mathcal{BA}_{23}^{8}$ is generally not equal to its converse. In this way, each arbitrarily complex pattern is put down to a number of bipartite arrangements, and each bipartite arrangement is treated as equally important. By doing this, we lose sight of those properties which can only be constituted by an arrangement of more than two lines. Such a property, for instance, is the *in-between* relation: one line is in-between two others. Sometimes two arrangements with a different number of lines may even constitute the same property. A line may be contained within a region which is made up of a number of other lines, perhaps a region being formed by only three lines, or possibly a region made up by an arrangement of many lines. These examples show that one bipartite arrangement cannot necessarily be treated on a par with any other bipartite arrangement. Instead it is important to identify those relations which together act upon the same property — two lines being arranged such that a third line can be put in-between them, or a number of lines being arranged to form the boundary of a region; these lines are to distinguish from those lines which are either inside or outside that region.

Two main approaches to investigating relationships involving more than two lines have been proposed. (Schlieder, 1995b) uses neighbourhood graphs as defined by (Freksa, 1992a) for describing changes in line segment relations. If transformations are continuous in space and time it will not be possible to produce an arbitrary sequence of line segment relations. In a continuous world each relation can only be followed by certain other relations, that is, by its conceptual neighbours. We defined this neighbourhood graph for the $\mathcal{BA}_{23}^{8}$-relations in Figure 3.11 and in Figure 3.12. By contrast, (Moratz et al., 2000) do not use a neighbourhood graph but instead define a relation algebra in order to be able to reason about line segment relations. We will also define a relation algebra with the relations of $\mathcal{BA}_{23}^{8}$ in the next section. It is then possible to apply constraint based reasoning techniques to line arrangements in order, for example, to test whether a particular set of $\mathcal{BA}_{23}^{8}$-relations gives a consistent scenario.

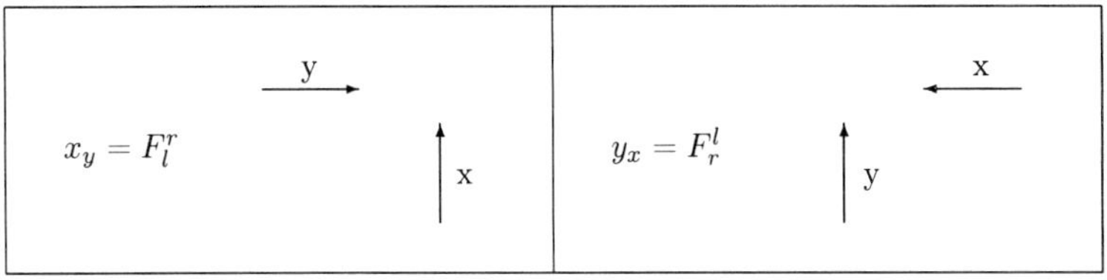

Figure 3.14: Example for the converse relation

3.2.2 A relation algebra on line arrangements

Considering (Ladkin & Maddux, 1994) a relation algebra is a nine-tuple:

$$\mathfrak{A} = (\mathcal{BA}_{23}^8, \cup, \cap, \bar{\ }, 0, 1, \circ, \breve{\ }, Id) \tag{3.3}$$

where $(\mathcal{BA}_{23}^8, \cup, \cap, \bar{\ }, 0, 1)$ is a Boolean algebra; $\mathcal{BA}_{23}^8$ is the universe, $\cup$ the union, $\cap$ the intersection, $\bar{\ }$ the complement, 0 is the empty relation, and 1 the universal relation; $\circ$ is a binary operation called the composition, $\breve{\ }$ is a unary operation called the converse, and $Id \in \mathcal{BA}_{23}^8$ is the identity relation. As the operations coincide with the usual set-theoretic operations on the relations in $\mathcal{BA}_{23}^8$, and since the universe is a set of binary relations, we obtain a proper relation algebra.

For any two binary relations R, $S \in \mathcal{BA}_{23}^8$, $R \cap S$ is the intersection of R and S, $R \cup S$ is the union of R and S, $R \circ S$ is the composition of R and S, $\breve{R}$ is the converse of R, and $\overline{R}$ is the complement of R. For intervals x, y, $z \in \mathbb{R}^2$, these operations are defined as:

$$\overline{R} = \{(x,y)|(x,y) \notin R\} \tag{3.4}$$
$$\breve{R} = \{(x,y)|(y,x) \in R\} \tag{3.5}$$
$$R \circ S = \{(x,z)|\exists y : (x,y) \in R \wedge (y,z) \in S\} \tag{3.6}$$
$$R \cap S = \{(x,y)|(x,y) \in R \wedge (x,y) \in S\} \tag{3.7}$$
$$R \cup S = \{(x,y)|(x,y) \in R \vee (x,y) \in S\} \tag{3.8}$$

In expressions without parentheses, the unary operations, i.e. converse and complement, are to be computed first, followed by composition, intersection, and union, in that order; repeated binary operations at the same priority level are to be computed from left to right. The compositions and converse relations are computed based on the semantics of the relations, and these two operations will now be discussed further.

Identifying the converse of a relation is a particularly useful operation: if x_y describes y with respect to x then for the converse relation it holds that $\breve{x}_y = y_x$. Figs. 3.16 - 3.17 show the table of converse relations, Φ denoting the position and ϕ the orientation. For instance, for the arrangement in Figure 3.14 it holds

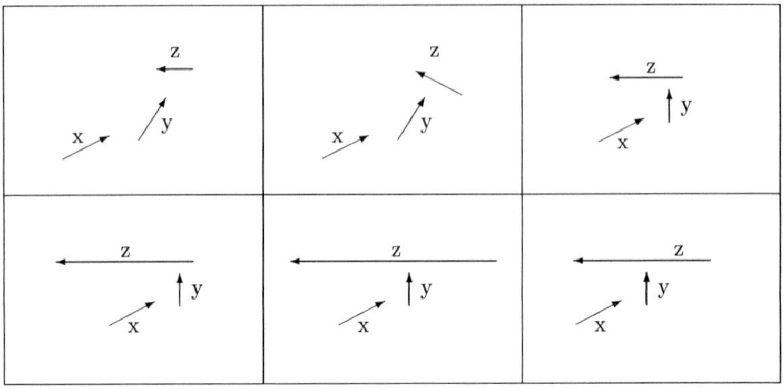

Figure 3.15: Example for the composition operation: given $x_y = F_m^{F_l}$ and $y_z = F_m$ there exist six possible arrangements for x_z

$x_y^\Phi =$ 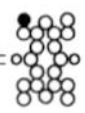 $\wedge\ x_y^\phi = \ominus \rightarrow y_x^\Phi =$ 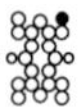 $\wedge\ y_x^\phi = \ominus$ (3.9)

(first column, second row in Figure 3.16). Note that Figure 3.16 contains some empty sets which denote impossible combinations of position and orientation.

Given x_y and y_z the relation x_z can be looked up in the composition table in which all possible transitivity relations are stored. Figs. 3.19 - 3.22 show the composition table. For instance,

$x_y =$ 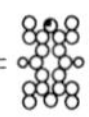 $\wedge\ y_z =$ 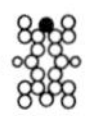 $\longrightarrow\ x_z =$ 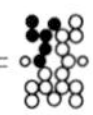 (3.10)

(second column, second row in the composition table in Figure 3.19). For the composition, we write $x_z = x_y \circ y_z$. The resulting orientation of z with respect to x is computed according to Figure 3.18. Figure 3.15 shows this situation. Given $x_y = F_m^{F_l}$ and $y_z = F_m$ there exist six possible arrangements for x_z, namely $\{F_l, F_m, FO_l, FO_{ml}, FC_l, C_l\}$.

We shall now consider the composition operation more thoroughly. The 125 different basic relations of $\mathcal{BA}_{23}^8$ allow $125^2 = 15625$ compositions. The orientation of y_z is recorded as y_z^ϕ, and its position as y_z^Φ. y_z^ϕ is irrelevant to the deduction of the position of z with respect to x, since x_y^ϕ, x_y^Φ, and y_z^Φ sufficiently constrain the position of z in relation to x; and the orientation of z with respect to x can be derived from the table in Figure 3.18. Hence different orientation variations of y_z can be pooled together in the composition table. Moreover, we can omit the identity relation since it behaves neutrally with respect to composition. Furthermore, all compositions with the empty relation are empty; and all compositions with the universal relation result in the universal relation,

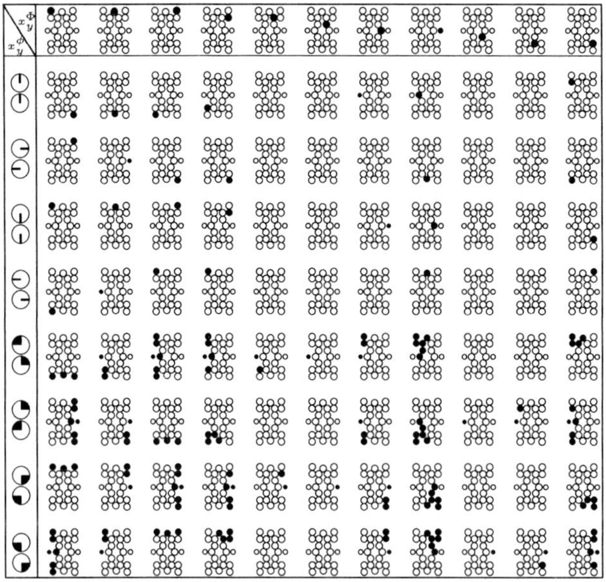

Figure 3.16: Converse relations: given x_y in the upper row, y_x is derived depending on the orientation of y with respect to x which determines the row; the orientation and the change in orientation is given in the first column

except for the composition with the empty relation. As a consequence, there are only 22 possible relations to be considered for y_z, leading to $124 * 22 = 2728$ entries in the composition table. Exploiting the available symmetries we can reduce the composition table further, to $31 * 22 = 682$ entries. We shall refer to the composition table as $\mathcal{CT}$. The following algorithm shows the computation of the 2728 compositions.

```
get-x_z (x_y, y_z)
    if x_y ∈ CT
        return CT(x_y, y_z)
    else if (x_y^φ)^-1 ∈ CT
        return CT((x_y^φ)^-1, (y_z^Φ)^-1)
    else
        return get-x_z(x_y^-1, y_z)^-1
```

$\mathcal{CT}(x_y, y_z)$ may simply be looked up in the composition table. $(x_y^{\phi})^{-1}$ con-

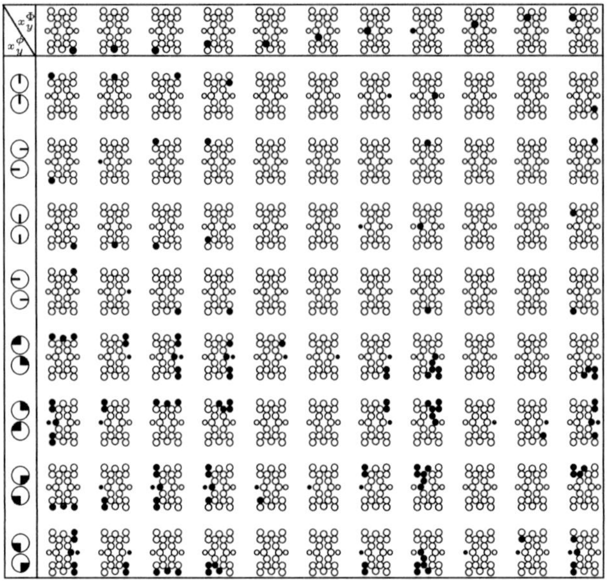

Figure 3.17: Converse relations — continued

siders the inverse of the orientation of y with respect to x while keeping the position the same; the inverse of the orientation is a rotation by 180°; $(y_z^\Phi)^{-1}$ can be expressed as $((y_z + 11)\ mod\ 22)$; this refers to the column in the composition table which is addressed by adding 11 columns to the column denoted by y_z; the modulo operator causing a return to the first column after the last column has been reached — this corresponds to a change in location by 180°; x_y^{-1} simultaneously takes the inverse of position and orientation, i.e. all occurrences of B are changed into F and vice versa, and all occurrences of l are changed into r and vice versa. Accordingly in the dimension parallel to the reference interval it holds that

$$B^{-1} = F \tag{3.11}$$

$$F^{-1} = B \tag{3.12}$$

and similarly for the perpendicular dimension

$$l^{-1} = r \tag{3.13}$$

$$r^{-1} = l \tag{3.14}$$

By the inverse operation it is also possible to compute the converse relations which are described in Figure 3.17 by those relations in Figure 3.16, since these

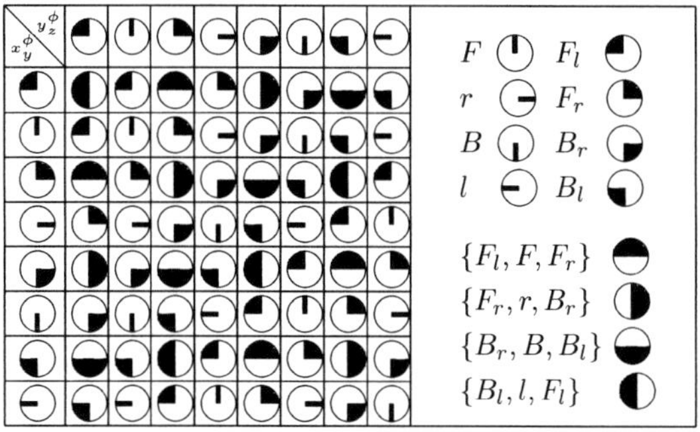

Figure 3.18: Given x_y^ϕ and y_z^ϕ possible orientations for x_z^ϕ are derived

two parts of the table of converse relations are inverse relative to each other. Furthermore, row-pairs 1 and 3, 2 and 4, 5 and 7, and 6 and 8 in Figure 3.16 are inverse relative to each other. Hence, there are ultimately only 4 rows and 11 columns to be considered, making a total of 44 cases which need to be explicitly defined in order to obtain all possible converse relations.

The composition algorithm terminates at the latest after the second recursion step depending on the given position and orientation of y with respect to x. Finally, the combinations of two orientations in order to compute the composition of the orientation $x_z^\phi = x_y^\phi \circ y_z^\phi$ are shown in Figure 3.18. By considering these orientation combinations, the previous composition algorithm, the empty relation, universal relation, and the identity relation, we obtain all 15625 compositions.

Two examples show the application of the composition algorithm:

1. $x_y = F_l^{B_l}$ and $y_z = FO_r^{F_l}$
   $\rightarrow x_y = F_l^{B_l} \notin \mathcal{CT}$ — $\mathcal{CT}$
   $\rightarrow (x_y^\phi)^{-1} = F_l^{F_r} \in \mathcal{CT}$ — (10th row)
   $\rightarrow \mathcal{CT}(F_l^{F_r}, FO_r^{-1}) = \mathcal{CT}(F_l^{F_r}, BO_l)$ — (4th column in Figure 3.21)
   $\rightarrow x_z^\Phi = \{F_l, FO_l, C_l\}$
   $\rightarrow x_z^\phi = B_l \circ F_l = \{B_r, B, B_l\}$ (Figure 3.18)
2. $x_y = B_m^{B_l}$ and $y_z = D_l^r$
   $\rightarrow x_y = B_m^{B_l} \notin \mathcal{CT}$
   $\rightarrow (x_y^\phi)^{-1} = B_m^{F_r} \notin \mathcal{CT}$ — $\mathcal{CT}$
   $\rightarrow\rightarrow x_y^{-1} = F_m^{F_r} \in \mathcal{CT}$ — (11th row)
   $\rightarrow\rightarrow x_z^{-1} = \mathcal{CT}(F_m^{F_r}, D_l) = \{F_l, F_m, F_r\}$ (8th column in Figure 3.21)
   $\rightarrow x_z^\Phi = (x_z^{-1})^{-1} = \{B_l, B_m, B_r\}$
   $\rightarrow x_z^\phi = B_l \circ r = F_l$ (Figure 3.18)

Though it is not necessary to use the orientation of z for the composition operation it can be used in order to restrict the number of possible results. This is because there may be some otherwise possible positional relations which are not consistent with the orientation of z, for instance, $x_z^{\Phi} = FC_l$ cannot hold if $x_z^{\phi} \in \{F_l, B_r\}$.

All the entries in the composition table have been verified manually. Examining the composition table we can make some observations:

(a) All compositions are valid.

(b) There are comparatively few inferences with unique results; most conclusions contain disjunctions of possible results, such disjunctions representing coarse information.

(c) Almost all disjunctions of alternative results form conceptual neighbourhoods. The only exceptions to this are disjunctions which are made up of different *contains* relations. These do not form a conceptual neighbourhood in every case since intersections are excluded; intersections would link different *contains* relations, such as FC_l and BC_r, into neighbourhoods; see, for instance, row 3 and column 5 in $\mathcal{CT}$ (Figure 3.19).

(d) In many cases neighbouring entries lead to similar neighbourhoods rather than to completely different neighbourhoods. But note that the composition table is not designed in such a way that all neighbouring preconditions form conceptual neighbours; for example, column seven, C_r, and column eight, D_r, do not form neighbouring concepts.

(e) Only a small fraction of the elements of the power set $\mathcal{P}(\mathcal{BA}_{23}^8)$ appear in the composition table.

(f) The composition table has entries only for atomic relations; for compound relations it is necessary to consider the unions of the compositions of the corresponding atomic relations.

(g) In the complete table, with 15625 compositions, there are several symmetries which we have used in order to reduce the number of entries necessary.

Since we are now equipped with a proper relation algebra it is possible to apply constraint based reasoning techniques. In the following, we consider a set $\mathbb{M}$ of n intervals in the plane. The satisfiability problem asks for the existence of n interval interpretations which are in accordance with a set of constraints Θ over the relations in $\mathcal{BA}_{23}^8$ which describe relations between the objects in $\mathbb{M}$. Such an interpretation does exist whenever all relations make up a consistent constraint network. As the identity relation holds for every interval by definition of $\mathcal{BA}_{23}^8$, such a constraint net is node consistent. The computational evaluation of the consistency of a constraint net is then performed as follows. In order to achieve arc-consistency

$$\forall x_y \in \Theta : x_y := x_y \cap \breve{y}_x \tag{3.15}$$

and in order to achieve path-consistency

$$\forall x_z \in \Theta : \forall y \in \mathbb{M} : x_z := \bigcap x_y \circ y_z \tag{3.16}$$

These two steps are to be performed until no new relations are inferred. As the empty relation is used to denote an inconsistent scenario, as soon as the empty relation is deduced, the constraint net will have been proven to be inconsistent because any empty relation will remain empty under any further computations of equations 3.15 and 3.16. When the network has stabilised without inferring the empty relation the constraint net has been shown to be consistent.

The procedure described can also be applied in order to integrate new knowledge. Each new relation extends the set Θ, and we update the constraint network by propagating such further constraints. Knowledge which is no longer valid can be considered by updating the constraint net after the corresponding relations in Θ have been deleted.

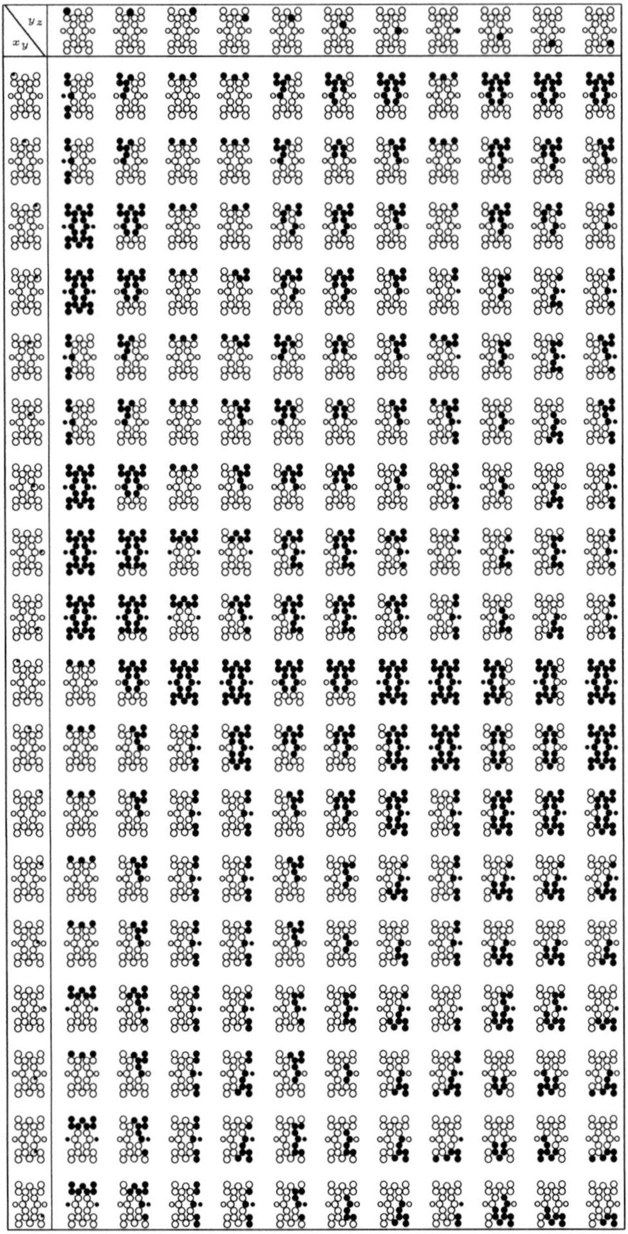

Figure 3.19: Composition table — part 1

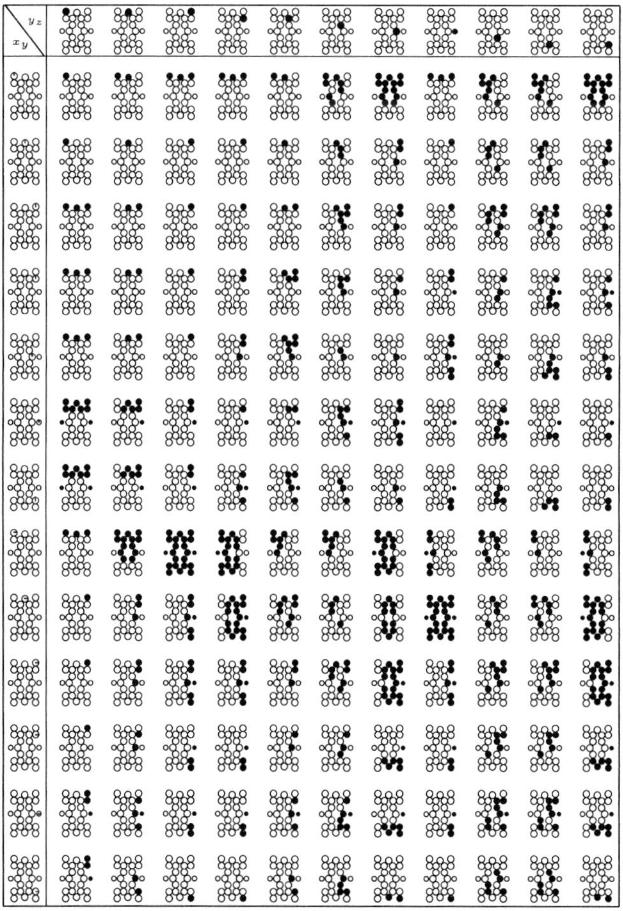

Figure 3.20: Composition table — part 1 continued

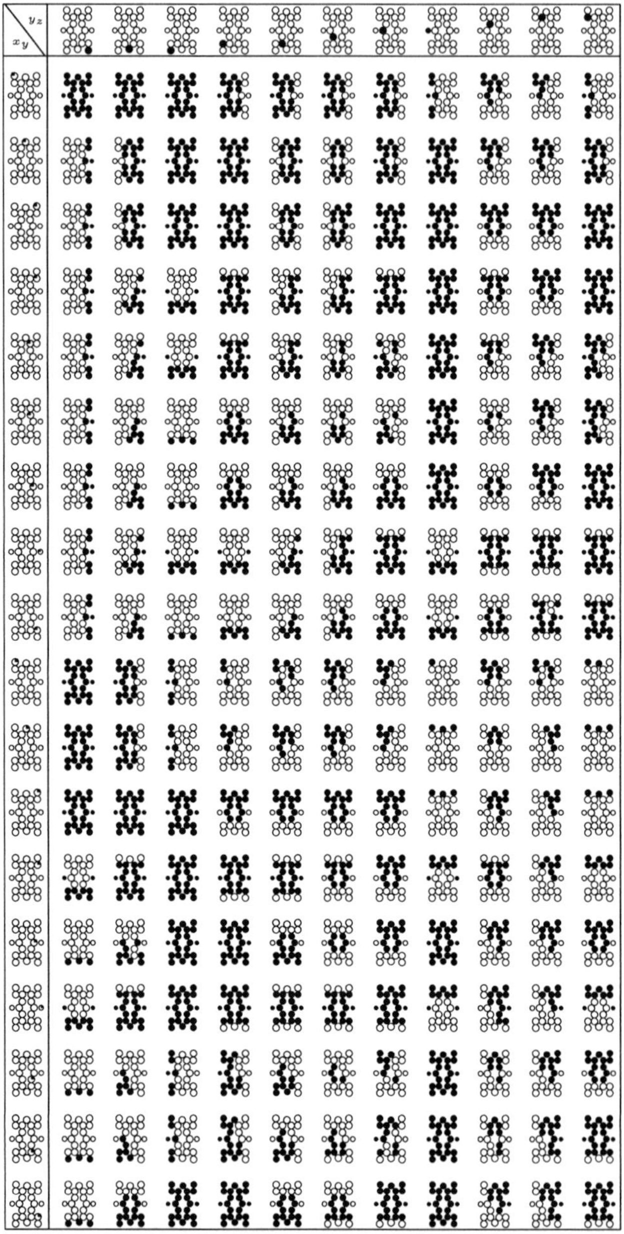

Figure 3.21: Composition table — part 2

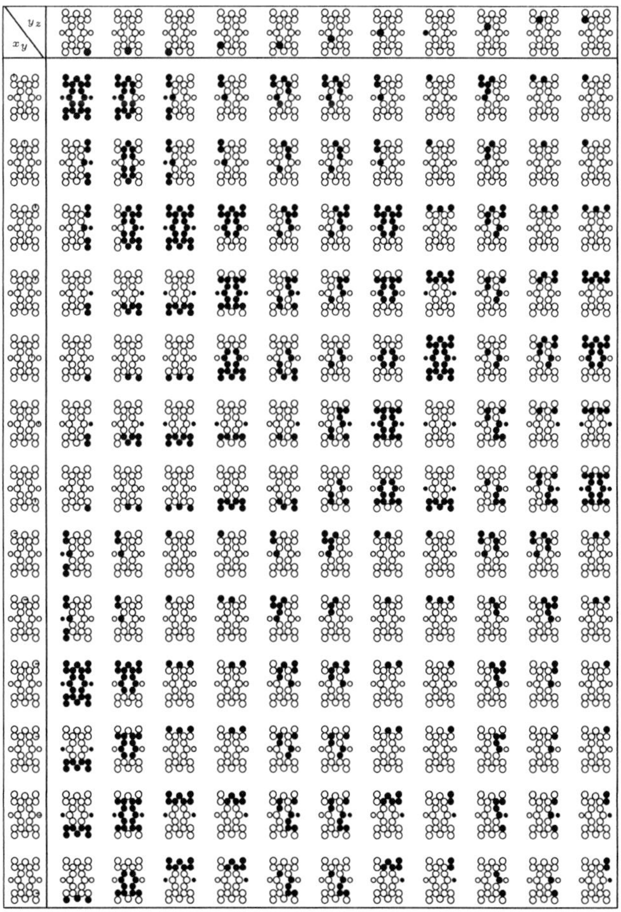

Figure 3.22: Composition table — part 2 continued

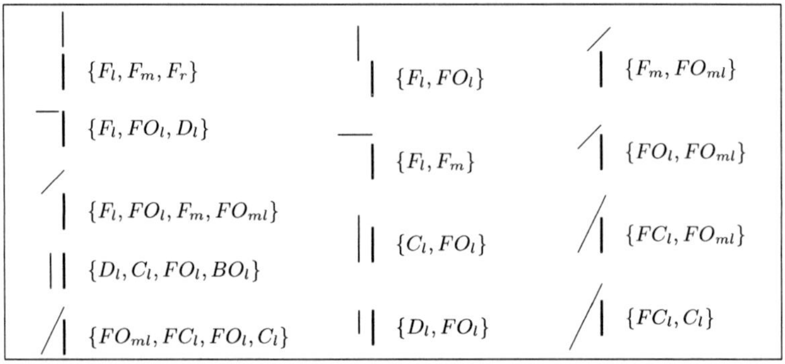

Figure 3.23: Singular relations defined by sets of general relations

3.2.3 Dealing with singularities and intersections

The relation algebra for line arrangements rests on a number of disconnection relations, all being in general position. Therefore, mechanisms are required in order to cope with singularities and intersections.

Singularities

In (Gottfried, 2004b) an appropriate method for dealing with singularities in qualitative representations is discussed. Whenever we encounter indeterminacy in the context of qualitative representations we consider sets of possible relations. This is not particularly precise, but precision is exactly what we want to avoid in qualitative reasoning. For example, when can we be sure whether parallel lines really are equal in length? Only when we have precise measuring tools. When working roughly, without such tools, there is always likely to be some uncertainty remaining (especially when sketching something roughly). At most we know that two lines in a given arrangement are *likely* to be equal in length, but at the same time we also know that they may be different to each other — very similar, but different. Similar relations form neighbourhoods in the $\mathcal{BA}_{23}$-graph, and such neighbourhoods circumscribe singular relations. Accordingly $\{D_l, C_l, FO_l, BO_l\}$ would seem to be quite an appropriate description of what we really know about two parallel lines which are *probably* equal in length.

Figure 3.23 shows how singular relations are represented by sets of general relations; note how easy it is to confuse the relations contained in one set when sketching these kinds of arrangement. Only a quarter of all relations are depicted, since the other relations are symmetrical to those in Figure 3.23. As with the disconnection relations of $\mathcal{BA}_{23}$, only disconnected singularities are considered. *Apparently connected singularities* are treated as *apparently con-*

nected general relations, i.e. they are conceived as disconnected relations in which distances become arbitrarily short.

Our knowledge can be said to get more uncertain near singular relations — this uncertainty being represented by sets comprising several possible relations rather than a single definite one. In particular, if two endpoints are in singular positions then these sets consist of three or four general relations, depending on whether the endpoints lie on the same singularity, e.g. $\{F_l, F_m, F_r\}$ in Figure 3.23, or on different singularities, e.g. $\{D_l, C_l, FO_l, BO_l\}$. By contrast, if there is only one endpoint in singular position the sets consist of only two general relations. We observe that all singularities are uniquely identified by this technique.

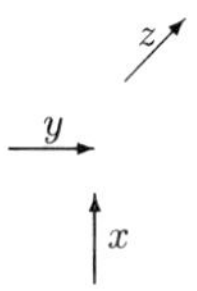

$$\begin{aligned} x_y &= \{F_l, F_m\} \\ y_z &= F_l \\ x_z &= x_y \circ y_z \\ &= \{F_l, F_m\} \circ F_l \\ &= F_l \circ F_l \cup F_m \circ F_l \\ &= \{F_l, F_m, F_r\} \cup \{F_r\} \\ &= \{F_l, F_m, F_r\} \end{aligned}$$

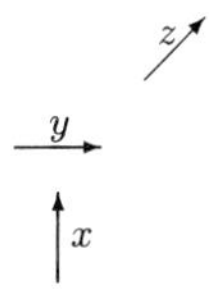

$$\begin{aligned} x_y &= F_m \\ y_z &= F_l \\ x_z &= x_y \circ y_z \\ &= F_m \circ F_l \\ &= F_r \end{aligned}$$

Figure 3.24: Transitivity with a singular relation (left), and without any singularity (right)

How does this representation of singular relations affects reasoning processes? Let us consider the example in Figure 3.24. We assume that we know the relations between x and y as well as those between y and z. Our goal is to infer the relationship between z and x, i.e. x_z. We do this by the composition operation. The left hand side of Figure 3.24 shows x_y in singular relation; the composition result is indeterminate. In comparison, the right hand side of Figure 3.24 shows x_y in general relation; here the composition result is less indeterminate. Note that this sort of reasoning is only necessary if x_z cannot be perceived directly; in Figure 3.24 the relation could of course be observed, rather than inferred.

Intersections

Having introduced a set of intersection-free basic relations, it is necessary to show how it is possible to deal with intersections using these. In pictorial space intersections are impossible, in perceptual space they are possible. Hence, from the viewpoint of pictorial space, an intersection must be considered a higher level concept. While two lines are perceived to intersect, in pictorial space four lines meet at a point, as demonstrated in Figure 3.25. This Figure shows an example of what we normally regard as an intersection. Such a case can be described by one uninterrupted line segment and two coincident line segments, or by two pairs of coincident line segments. Additionally, there is no gap between two coincident line segments, i.e. two coincident line segments meet.

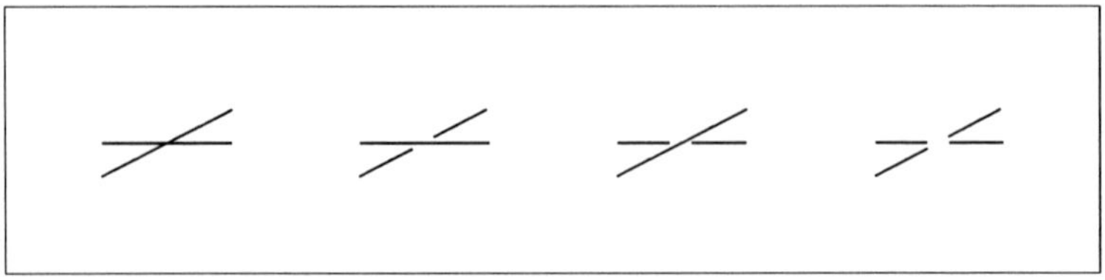

Figure 3.25: How many line segments are there?

In order to describe this situation in pictorial space we need a concept which distinguishes whether two disjoint lines meet or whether they are separated from each other. Note that two lines which meet remain disjoint because they can be removed separately — they are not connected. For two lines x and y which meet, we write $x\|y$, and Figure 3.26 shows how it is possible to define by $\mathcal{BA}_{23}^{8}$ that two lines coincide. Two coincident line segments are oriented identically, and one of them is in the front-middle with respect to the other one. Denominating them as x and y, it holds that $x_y = F_m^F$. The positional relation F_m ensures that one interval is somewhere in the front-middle of the other one (left of Figure 3.26); the orientation relation ensures that both intervals point in the same direction (middle of Figure 3.26). Taking positional relation and orientation relation together the coincidence of these line segments is completely specified (right of Figure 3.26). Of course, the two lines remain coincident whether the primary line is placed before or after the reference line, and also if its orientation is reversed (i.e. changed by 180°) relative to the reference line.

The concept we refer to as an intersection can then be defined on the basis of disjoint line arrangements in the plane: two lines in perceptual space intersect if in pictorial space there are four lines, v, w, x, and y, for which it holds that

$$v_w = \{F_m, B_m\}^{\{F,B\}} \wedge v\|w \wedge$$

$$x_y = \{F_m, B_m\}^{\{F,B\}} \wedge x\|y \wedge$$

$$v_x = D_l \wedge v_y = F_r \wedge w_x = B_l \wedge w_y = D_r \qquad (3.17)$$

In perceptual space v and w form one single line, and so do x and y — compare Figure 3.25. The singularities between v_x, v_y, w_x, and w_y are treated in accordance with section 3.1.3. (In the case of a cross, relations such as $v_x = F_l$ and $v_y = F_r$ may also be appropriate — it is only important that x is left of v while y is right of v.)

Note that a situation such as the one on the right hand side of Figure 3.26 would be circumscribed by a set of possible general relations, namely $\{F_l, F_m, F_r\}$, when interpreting an image, as explained earlier. But in order

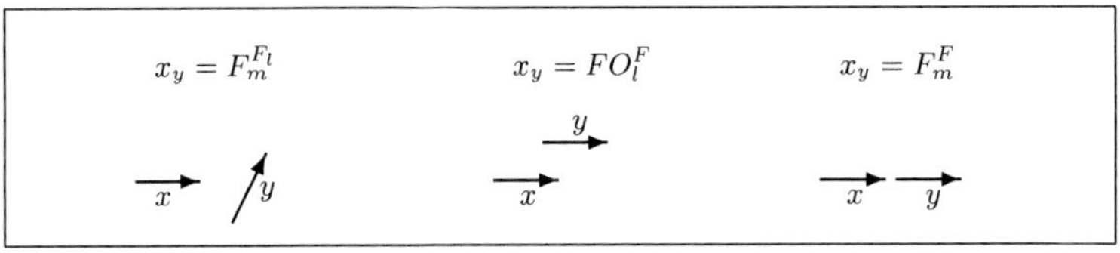

Figure 3.26: Defining coincidence by $\mathcal{BA}_{23}^8$: on the left y is in relation F_m with respect to x but oriented differently; in the middle y is not in relation F_m but is oriented identically to x, and on the right y is again in relation F_m with respect to x and additionally x and y have the same orientation

to *model* a special situation it is useful to constrain such an arrangement more precisely, in this case by F_m^F. This holds for all those singularities where both endpoints of the primary line lie on the same singularity — this distinguishes $\mathcal{BA}_{125}$ from $\mathcal{BA}_{113}$ and relates to the distinction between indeterminacy in perception and the requirement for accuracy when modelling a situation.

Chapter 4

Characterising polygons qualitatively

So far, we have analysed arrangements between disconnected intervals. Intervals can be considered as abstractions of objects, representing their intrinsic orientation, their extension seen from a particular viewpoint, or their direction of movement. The length of an interval may represent the length of the object, its speed, or the distance between two objects or components of one object. However, in order to describe the shape of single objects we need to consider arrangements of connected lines. That is, we are interested in describing the outlines of shapes, which constrain how objects are related to other objects and their environment.

We have previously referred to Attneave's investigations on the subject of using polygons for approximating shapes (chapter 2.1.4). Outlines of shapes can be usefully approximated by polygons, which form a special class of line arrangements, and we will therefore consider the characterisation of polygons as a reasonable alternative to the characterisation of arbitrary shapes. The term *polygon* derives from the Greek *poly* (many) and *gonia* (angle). A polygon is a planar path composed of a finite number of sequential straight line segments. Sometimes the term polygon also refers to the interior of a polygon, but in the present context, it will be used only to indicate a group of straight line segments which either form a closed path or an open one. More formally, we define polygons as follows:

Definition 4.1 (Polygon) *A polygon P is an n-tuple* $(x_1, x_2, .., x_n), n \in \mathbb{N}$ *of line segments which are connected at their endpoints. One line,* x_1*, is distinguished as the first line segment, and another one,* x_n*, is distinguished as the last line segment. The orientation of line segment* x_i *is defined in such a way that its front connects to the back of* x_{i+1}.

In the following, line segments may be identified either with x_1, x_2, x_3, etc, or with x, y, z etc., n denotes the number of lines comprising the polygon under consideration, and $x \in P$ will denote that x is a line-component of polygon P. What distinguishes polygons from arbitrary line arrangements is that each polygon defines an order for its lines. This order can be described by the definition of the successor of a line:

Definition 4.2 (Successor) *Each line $x_i, i = 1, .., n-1$ of a polygon P has exactly one successor $x'_i = x_{i+1}$. For closed polygons it holds that $x'_n = x_1$.*

In order to refer to indirect successors we use also symbols such as $<$ and $\leq$. It is quite worthwhile to investigate polygons more thoroughly, since they have often been characterised using quite general properties — prominent among them:

1. A polygon is either *open* or *closed* depending on whether each line segment, including x_n, has a successor. A closed polygon forms a path that, followed from any point on the path, will return to the starting point after passing through every other point forming the polygon.
2. A polygon is *simple* if there are no two lines which intersect. Otherwise it is called *complex*.
3. A closed and simple polygon is *convex* if for two arbitrary points contained in the interior of the polygon it holds that the straight line which connects these points is wholly contained in the polygon. Otherwise the polygon is *concave*.
4. A convex polygon whose vertices all lie on its circumcircle is *concyclic* or a *cyclic polygon*.
5. A cyclic polygon is called *regular* if all its sides are of equal length and all its angles are equal. It is otherwise *irregular*.

These properties allow to distinguish some shapes. But as soon as more sophisticated shapes are to be distinguished such properties are too general. In fact, taking the polygonal approximation of the silhouette of an arbitrary object, we always obtain a closed and simple polygon, being in most instances concave, non-cyclic, and irregular. It is rather the question how to distinguish different polygons which are all closed, simple, concave, non-cyclic, and irregular; or, when dealing with trajectories, we have to distinguish polygons which are in general open and non-simple.

Eventually, polygons are named according to the number of sides involved, combining a Greek root with the suffix *-gon*, among others trigon = 3, tetragon = 4, pentagon = 5, hexagon = 6, heptagon = 7, octagon = 8, enneagon = 9, decagon = 10, and googolgon = 10^{100}. Frequently, the abbreviation n-gon is used to refer to a polygon made up of n lines. We argue that there is much more to say about polygons: they can be conceived as special arrangements of lines. We already analysed line arrangements and considered distinctions which are described by $\mathcal{BA}$, i.e. distinctions which are confined to disconnected lines. But besides disconnected parts polygons are distinguished by the way how lines are connected. Therefore, we shall first of all explicitly analyse which kinds of connection relations in polygons exist, and we shall see how these connected sub-polygons describe local polygonal properties. By contrast, we shall later learn how disconnected line arrangements describe global properties of polygons.

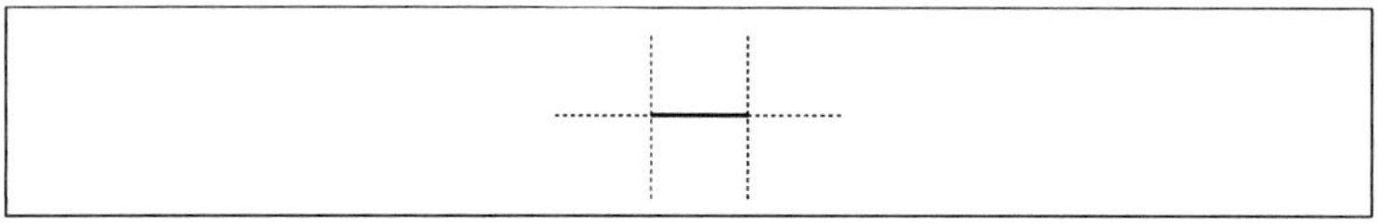

Figure 4.1: A reference line and resulting orientation grid

4.1 Local orientation

We will now discuss local polygonal properties. Specifically, we will investigate how the local orientation progression can be described by arrangements of connected lines. We will refer to descriptions of connected lines as to *line tracks* which define equivalence classes of polygons by emphasising specific polygonal properties. They belong to the class of shape descriptions which are known as *local feature schemes* (Meathrel & Galton, 2000) since they are obtained by determining relations between line segments which are adjacent in the outline of a shape.

How are line tracks related to bipartite arrangements? They can be conceived as a number of line segments with pairs of lines meeting at their endpoints, i.e. we are concerned with a special subset of *meet*-relations in two dimensions. From the point of view of $\mathcal{BA}_{23}$ such relations involve singularities, which are described by subsets of $\mathcal{BA}_{23}$-relations which act to circumscribe meet-relations. Line track relations are fundamental when characterising polygons Therefore, we will systematically enumerate them and represent them in the form of shape primitives, instead of considering them as singularities.

4.1.1 Bipartite line tracks

The most general line tracks are made up of two lines. They describe obtuse and acute angles and are represented by $\mathcal{BA}(28)$ and $\mathcal{BA}(30)$. In order to construct shape primitives we consider only general positions, as for $\mathcal{BA}_{23}$. Since there are six general positions with respect to a reference line (as shown in Figure 4.1) we obtain six bipartite relations, which are depicted in Figure 4.2. The line a-b is the reference line, and c is the endpoint of the related line, which varies from relation to relation. These primitives have been introduced in (Gottfried, 2003a) where they are referred to as *bipartite line tracks*, since they consist of two lines. Bipartite line tracks ($\mathcal{BLT}$, for short) represent six different meet-relations in two dimensions. The i-th relation can be indicated by $\mathcal{BLT}(i)$, for $i \in \{1, 2, 3, 4, 5, 6\}$.

Bipartite line tracks are closely related to another local feature scheme which also provides a description of polygons: Jungert (1993) considers two adjacent lines in an oriented polygon in order to characterise a vertex which connects these lines. This description is obtained by *symbolic slope projections*, that is,

$\mathcal{BLT}(1)\equiv\mathcal{BA}(30)$	$\mathcal{BLT}(2)\equiv\mathcal{BA}(60)$	$\mathcal{BLT}(3)\equiv\mathcal{BA}(90)$	$\mathcal{BLT}(4)\equiv\mathcal{BA}(120)$	$\mathcal{BLT}(5)\equiv\mathcal{BA}(150)$	$\mathcal{BLT}(6)\equiv\mathcal{BA}(180)$

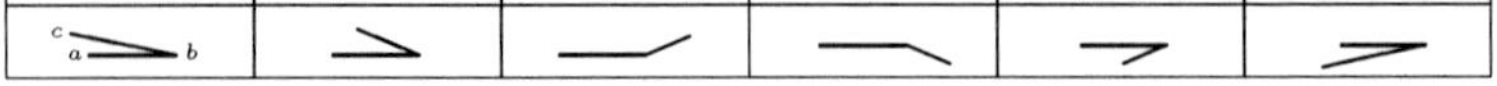

Figure 4.2: Six distinguishable classes of bipartite line tracks; as indicated by the $\mathcal{BA}$-number, these line tracks correspond to special bipartite arrangements

the endpoint, p_{j+2}, of the second line is projected (parallel to the first line; see Figure 4.3) onto the x-axis. From this, we can derive whether the vertex at p_{j+1} forms a convex or concave part of the polygon. We can see from the projection string (the ordering of the projections on the x-axis — since p_j and p_{j+1} define the slope of the projection, they are projected onto the same point) whether p_{j+2} comes before p_j or whether it follows it on the x-axis. In the first case the vertex between the lines is concave, otherwise it is convex. This is based on Jungert's stipulation that a polygon is always to be traversed anticlockwise, the figure being left of the polygonal path, the ground being on its right. If we follow this convention then $\mathcal{BLT}(1)$, $\mathcal{BLT}(2)$, and $\mathcal{BLT}(3)$ correspond to convex parts and $\mathcal{BLT}(4)$, $\mathcal{BLT}(5)$, and $\mathcal{BLT}(6)$ form concave parts. Note that Jungert has to specify explicitly the special case in which the first line runs parallel to the x-axis, because the slope of the first line is then zero, i.e. no slope projection to the x-axis exists. Furthermore, he has to distinguish whether the slope of the first line increases or decreases.

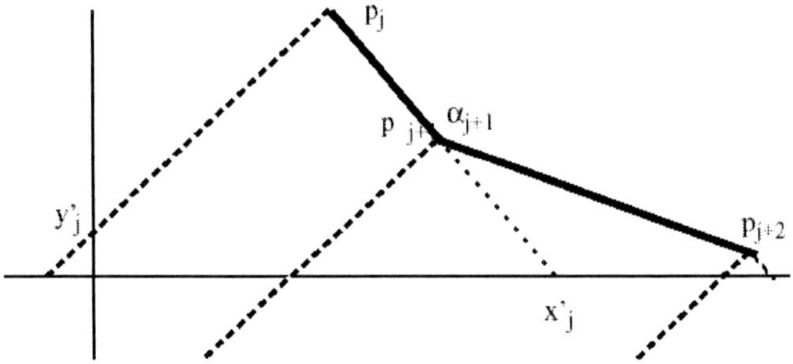

Figure 4.3: Two adjacent lines $p_j - p_{j+1}$ and $p_{j+1} - p_{j+2}$ and the dotted projection lines — from (Jungert, 1993)

In order to distinguish between acute and obtuse angles, the projections on the y-axis must be considered. These are obtained by projecting the points perpendicular to the first slope onto the y-axis. If p_{j+1} and p_{j+2} are equal regarding this projection a right angle can be inferred; if $p_{j+1} < p_{j+2}$ then the angle is obtuse otherwise it is acute. Again, several different cases must be distinguished, depending on the orientation of the projection line with respect to the plane, i.e. with respect to the axis of ordinates and the axis of abscissae. For bipartite line tracks it holds that the angles of $\mathcal{BLT}(1)$, $\mathcal{BLT}(2)$, $\mathcal{BLT}(5)$, and $\mathcal{BLT}(6)$ are acute, those of $\mathcal{BLT}(3)$ and $\mathcal{BLT}(4)$ are obtuse.

Polygons

Jungert traverses around a polygon in order to determine for each vertex its properties as just described. So do we. As a consequence a closed polygon with $k \geq 2$ lines is described as a vector of k $\mathcal{BLT}$s, an open polygon as a vector of $k-1$ $\mathcal{BLT}$s: $\mathcal{BLT}(i,j,...), i,j \in \{1,2,3,4,5,6\}$. In order to be able to treat each arbitrary polygon with k lines, and not only those which consist of a multiple of two lines, a vector of $\mathcal{BLT}$s describes a polygon in such a way that two consecutive $\mathcal{BLT}$s share one line, i.e. each line functions as both as a primary line and as the reference line in the context of the successive $\mathcal{BLT}$. A $\mathcal{BLT}$ can be regarded as a property of an arrangement of two connected lines, or as a property of a vertex, as Jungert does. Since any two connected lines, rather than just the point which connects those lines, determine the $\mathcal{BLT}$ (and, identically, the kind of vertex in Jungert's system) it seems to be more appropriate to regard it as a property of a specific line arrangement and not of a single point.

Jungert stipulates that any polygon is to be analysed beginning at the uppermost left corner of the polygon. We do not want to commit ourselves to any external reference system and therefore a $\mathcal{BLT}$ description of a closed polygon can begin with any line. As a consequence, for closed polygons, different $\mathcal{BLT}$ descriptions are obtained depending on where one starts to traverse the polygon. However, any description can be converted into another equivalent description by means of a cyclic permutation of the $\mathcal{BLT}$s involved. We can then choose for any given polygon the description that comes first in the ordering with respect to the $\mathcal{BLT}$ numbers. Figure 4.4 shows an example.

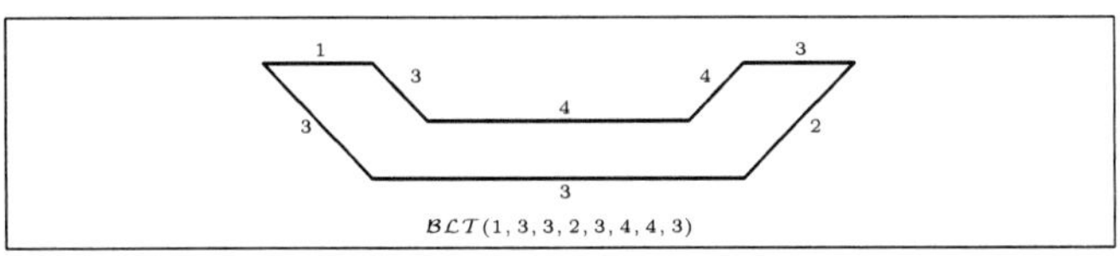

Figure 4.4: Example polygon described (anticlockwise) by bipartite line tracks

Neighbourhood graph

While Jungert's approach consists of describing the process which leads to the characterisation of polygons, we have made explicit a representation with six different relations. Distinguishing between a representation and the process involved in obtaining a description which is based on this representation is important. Once in possession of an explicit representation, it can be possible to carefully exploit further characteristics of that representation. In our case, such characteristics are found by means of the neighbourhood graph. Figure 4.5 shows the neighbourhood graph of $\mathcal{BLT}$-relations. As with the neighbourhood graph of $\mathcal{BA}_{23}$, two relations are neighbours if they can be transformed into one another by continuously moving one endpoint to another position whilst crossing a line of the reference system (Figure 4.1) exactly once. With the aid of the neighbourhood graph it is possible to define similarities between polygons and to describe possible deformations (Gottfried, 2003a). Furthermore, it allows us to deal with singular positions, as we shall learn in the next paragraph.

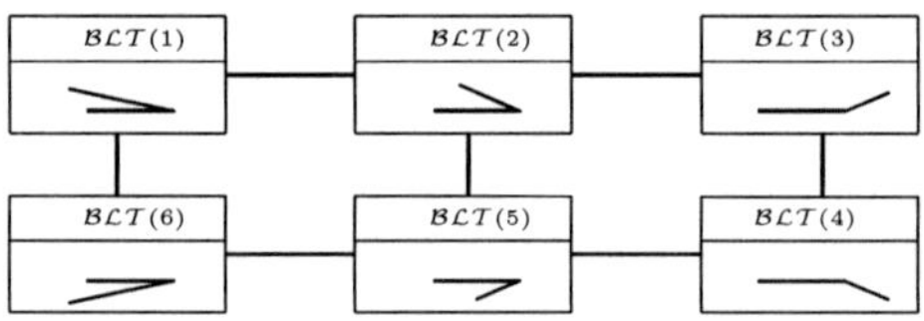

Figure 4.5: Neighbourhood graph of the $\mathcal{BLT}$-relations

Singular line tracks

Those $\mathcal{BLT}$s which have been considered so far would have been better denominated as *general bipartite line tracks* because singular positions were disregarded. Line tracks which consist entirely of singular positions are of special interest with respect to the boundaries of artificial objects which often have sides that are perpendicular to each other. Such sides form right angles, i.e. we are neither concerned with acute nor with obtuse angles but with line arrangements in singular positions. A *singular bipartite line track* is referred to as $\mathcal{BLT}(a\text{-}b)$, i.e. $\mathcal{BLT}(a\text{-}b)$ refers to an edge in the neighbourhood graph, where a and b are the relations which are connected by this edge. As an example, consider $\mathcal{BLT}(1\text{-}2)$ in Figure 4.6. This singular position is in between the two general positions $\mathcal{BLT}(1)$ and $\mathcal{BLT}(2)$.

Singular line tracks with the endpoints lying on the positions 2, 5, 8, 11, or 14 (see Figure 3.5) correspond to degenerated cases where we only perceive a single line. These positions are denoted in the following way: $\mathcal{BLT}(3\text{–}4)$ (position 14 in Figure 3.5), $\mathcal{BLT}$(2–3–4–5) (position 11), $\mathcal{BLT}$(2–5) (position 8), $\mathcal{BLT}$(1–2–5–6) (position 5), and $\mathcal{BLT}$(1–6) (position 2). This notation allows us to indicate all those general positions between which there is uncertainty regarding the position of a point or line. Uncertainty arises whenever a position is perceived as being somewhere near the transition between two or more general positions. In most cases, a singular position will not definitely be recognised as being singular. One will simply have doubts regarding a general position which lies near the transition to another general position. Thus, the way in which singular positions are treated is related to the question of dealing with uncertainty about positions near boundaries of two or more neighbouring regions. Accordingly, in (Gottfried, 2004b) singular relations are represented by sets of general relations. This is justified by the fact that singular relations correspond to situations about which a perceptual system is uncertain, and in qualitative reasoning uncertainty is frequently dealt with by considering sets of possible situations. In this way singular relations are second-order relations which are circumscribed by basic relations rather than being basic relations themselves. By contrast, Jungert treats arrangements in singular position at the same level as those in general position. Also, he distinguishes $\mathcal{BLT}$(2-3) and $\mathcal{BLT}$(4-5), but not $\mathcal{BLT}$(1-2) and $\mathcal{BLT}$(5-6).

Non-oriented primitives

With bipartite line tracks we introduced a shape description which consists of six oriented shape primitives. Oriented primitives allow us to discriminate contour parts which are oriented to the shape of an object from those which are mirror-symmetrical to the former parts, i.e. parts which are oriented to the background. Hence, we distinguish $\mathcal{BLT}$(1) and $\mathcal{BLT}$(6), $\mathcal{BLT}$(2) and $\mathcal{BLT}$(5), as well as $\mathcal{BLT}$(3) and $\mathcal{BLT}$(4) (see Figure 4.5). In ambiguous cases, $\mathcal{BLT}_6$ is used to refer to these six relations.

In the context of vision and in particular when being faced with imprecise sketches, we often have to deal with incomplete shape information. As a consequence, partial shape information has to be described. But frequently nothing can be said concerning the orientation of shape parts when it is not known on which side of any contour-part the figure or the ground is. In these cases we

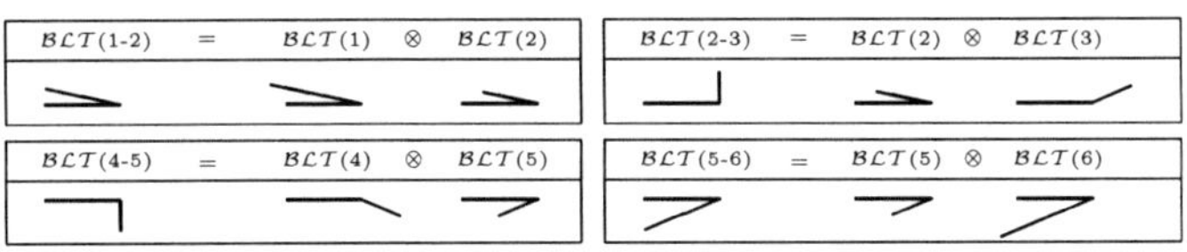

Figure 4.6: Singular bipartite line tracks

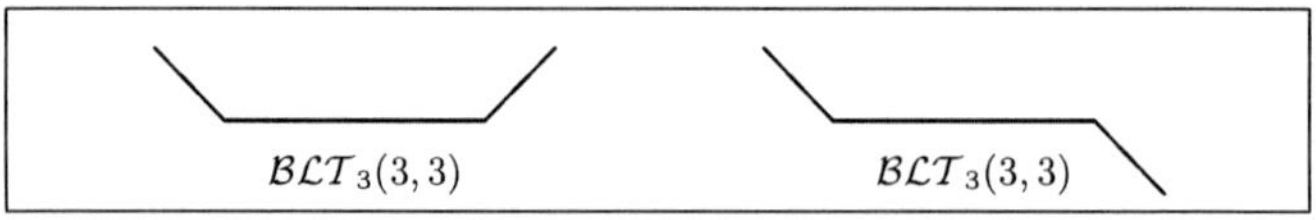

Figure 4.7: Two different polygons not distinguishable by $\mathcal{BLT}_3$

have to consider primitives which are not oriented. Putting together the symmetrical relations, namely $\mathcal{BLT}(1)$ and $\mathcal{BLT}(6)$, $\mathcal{BLT}(2)$ and $\mathcal{BLT}(5)$, as well as $\mathcal{BLT}(3)$ and $\mathcal{BLT}(4)$, we obtain three non-oriented relations, i.e. one obtuse angled primitive, like $\mathcal{BLT}_6(3)$, and two kinds of acute angled primitives, like $\mathcal{BLT}_6(1)$ and $\mathcal{BLT}_6(2)$. We shall refer to these relations as to $\mathcal{BLT}_3$. Such non-oriented primitives have the disadvantage that some shapes can no longer be distinguished; the two polygonal parts in Figure 4.7, for example. In order to distinguish more complex non-oriented primitives we will consider line tracks made up of three lines in the next section. But before that, we shall have a look at some combinations of $\mathcal{BLT}$s, in order to demonstrate of what kinds of polygons can be distinguished, even by non-oriented $\mathcal{BLT}$s.

Figure 4.8 depicts polygons which are made up of $\mathcal{BLT}_3(2)$ and $\mathcal{BLT}_3(3)$. Only the distinction between acute and obtuse angles is considered, i.e. $\mathcal{BLT}_3(1)$ and $\mathcal{BLT}_3(2)$ are treated equally. We refer to these relations as $\mathcal{BLT}_2$. In addition, rectangular angles ($\mathcal{BLT}_3$(2-3)) are considered. We are able to distinguish a considerable number of different quadrilaterals, though $\mathcal{BLT}_2$ is only based on the distinction between acute and obtuse angles. Thus, $\mathcal{BLT}_2$ is both simple and expressive, and we will therefore consider one further set of $\mathcal{BLT}$ relations. This set, which we will call $\mathcal{BLT}_4$, is the oriented counterpart of $\mathcal{BLT}_2$, as $\mathcal{BLT}_6$ is the oriented counterpart of $\mathcal{BLT}_3$.

Summary

While Jungert's approach requires the slope projections onto the axis of ordinates and the axis of abscissae, $\mathcal{BLT}$s are defined in a way which is independent of any external reference system. There is no need for such a reference system since the orientation grid suffices for the distinction of acute, right, and obtuse angles, and for the distinction of convex and concave shapes as well as for determining where the concavities of a polygon are situated. Where an external reference system is used, a number of different cases must be considered in order to obtain these properties and this makes Jungert's approach rather complex. On the other hand, a further possibility when using an external reference system is the determination of *extreme points*. These describe whether a vertex is north of the other two vertices to which it is adjacent, or whether it is south, west, or east of them, or between them with respect to both axes. This is achieved by

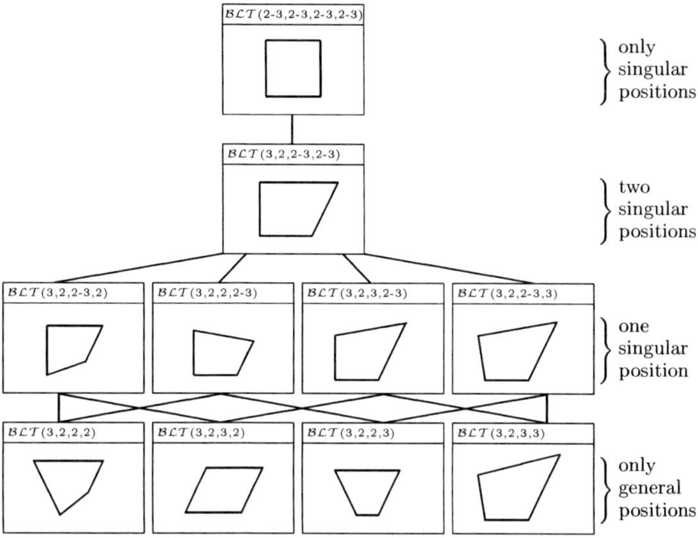

Figure 4.8: Simple convex quadrilaterals described by $\mathcal{BLT}_2$ arranged in a neighbourhood graph; the anticlockwise encoding starts always with the bottom line

orthogonally projecting the vertices onto the axis of ordinates and the axis of abscissae. Unfortunately, when using such extreme points Jungert's description ceases to be rotation invariant.

By contrast to Jungert, we distinguish between the construction process of the representation and the representation itself. We have produced a set of explicit primitives ($\mathcal{BLT}_4$) which allow for the same distinctions as Jungert's system. There are only two exceptions: the rotation-variant extreme points and the consideration of an ordering between adjacent angles. The latter describes which of two adjacent angles is larger, resulting in a list of *smaller than* and *larger than* relations, for all angles around the polygon. We could easily supplement a $\mathcal{BLT}$ description with such relations if required, however. On the other hand, there are some properties which are outside the scope of Jungert's approach. Using the orientation grid there are relations which allow distinctions between different kinds of acute angles. These relations are contained in both $\mathcal{BLT}_3$ and $\mathcal{BLT}_6$. Non-oriented primitives are defined by $\mathcal{BLT}_2$ and $\mathcal{BLT}_3$. These make sense whenever fragments of shapes are to be analysed, as there is frequently a lack of knowledge about orientations in shape-fragments. More sophisticated non-oriented relations are introduced in the following section.

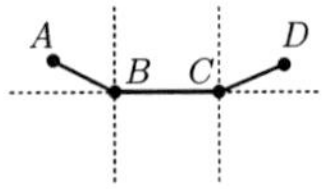

B and C are the reference points of the reference line.
A and D are the endpoints of the two related lines.

Figure 4.9: A line track ($\overline{AB}, \overline{BC}, \overline{CD}$) which consists of three connected lines

4.1.2 Tripartite line tracks

When we allow for a third line in a line track, a more expressive set of line tracks is obtained. Three connected lines can be arranged in more different ways than two connected lines, since we must consider all possible combinations of two connected bipartite arrangements. Local orientation information is either described by a reference line and its successor, as for $\mathcal{BLT}$-relations, or by considering both the predecessor of the reference segment and its successor. In any case, local orientation information is defined by the way in which adjacent lines are connected to each other. In this section we investigate relations which are made up by three connected lines, i.e. we consider both the predecessor and successor of the reference segment. Figure 4.9 outlines the description of three connected lines using the reference system in Figure 4.1. The two endpoints of the medial line, B and C, define the reference line, the two outer lines are related to this reference line. Note that nothing is specified about the relationship between the two outer lines, since these lines are *not* connected. As such their relationship concerns non-local orientation information, which is dealt with later on.

$\mathcal{TLT}_{36}$

As Figure 4.9 shows, each of the two endpoints of a tripartite polygon can lie in one of six possible areas. Therefore, there are $6^2 = 36$ different relations, which are depicted in Figure 4.10. These line tracks have been introduced in (Gottfried, 2002), and, since they are composed of three lines, they are referred to as *tripartite line tracks* ($\mathcal{TLT}$). The i-th relation is indicated by $\mathcal{TLT}(i)$. The medial line is considered to be oriented, in Figure 4.10, from left to right with respect to the image plane.

Having earlier argued that non-oriented primitives are particularly useful when dealing with incomplete shapes, $\mathcal{TLT}$-classes which differ only regarding symmetrical variations can be put into a single class in order to obtain a set of non-oriented cases. Consider, for example, $\mathcal{TLT}_{36}(1)$, $\mathcal{TLT}_{36}(15)$, $\mathcal{TLT}_{36}(22)$, and $\mathcal{TLT}_{36}(36)$. These relations look quite similar since in each case the two endpoints are lying in the same area with respect to the reference line; only their relation to the orientation of the medial line differs. Similar symmetrical relationships hold for all other relations. Being interested in non-oriented shape

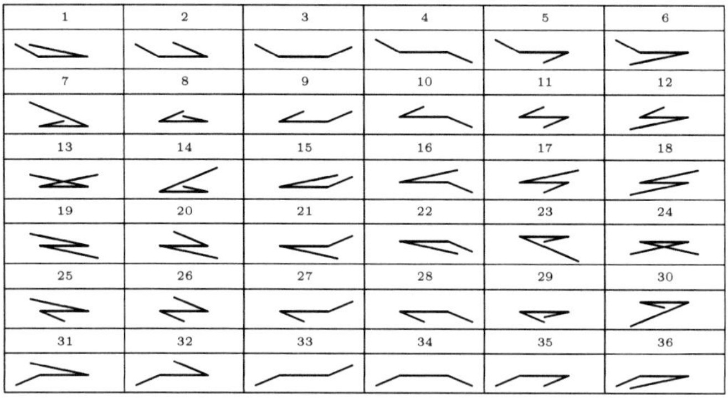

Figure 4.10: 36 distinguishable classes of oriented line track arrangements with three connected lines

primitives, in the next paragraph a set of $\mathcal{TLT}$ relations are considered in which the medial line is no longer oriented.

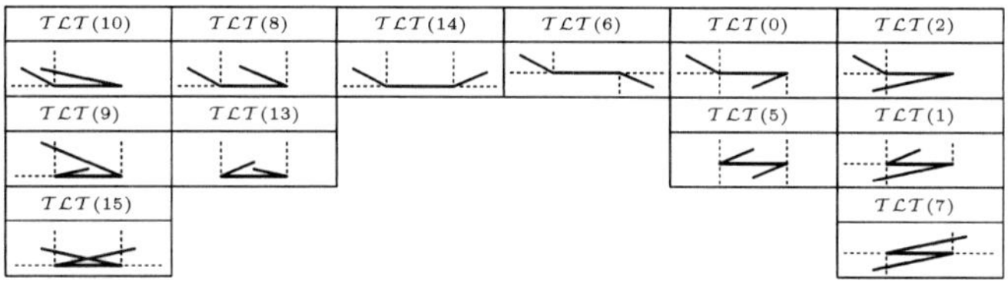

Figure 4.11: Twelve equivalence classes distinguished by *tripartite line tracks*

$\mathcal{TLT}_{12}$

Whereas Jungert is restricted to oriented shape parts, we are capable of dealing with these non-oriented $\mathcal{TLT}s$, which are depicted in Figure 4.11[1]. There are twelve of them in total, since symmetrical $\mathcal{TLT}_{36}$ relations are put into equivalence classes. We are now able to distinguish the two tripartite polygons in Figure 4.7 by $\mathcal{TLT}_{12}(6)$ and $\mathcal{TLT}_{12}(14)$.

A conceptual neighbourhood graph (as introduced for bipartite arrangements) is shown in Figure 4.12, and allows us to draw conclusions about the

[1]Note that in (Gottfried, 2002) an encoding has been introduced in order to directly refer to properties of $\mathcal{TLT}s$. We use those $\mathcal{TLT}$-numbers but will not discuss this encoding here.

relationships among the $\mathcal{TLT}$ relations, especially with regard to their similarities to each other. Singular tripartite line tracks are defined in the same way as singular bipartite line tracks. For example, $\mathcal{TLT}$(8-13), $\mathcal{TLT}$(8-14), and $\mathcal{TLT}$(13-8-14) are shown in Figure 4.13. The latter demonstrates a line track in which both sidelines are in singular position.

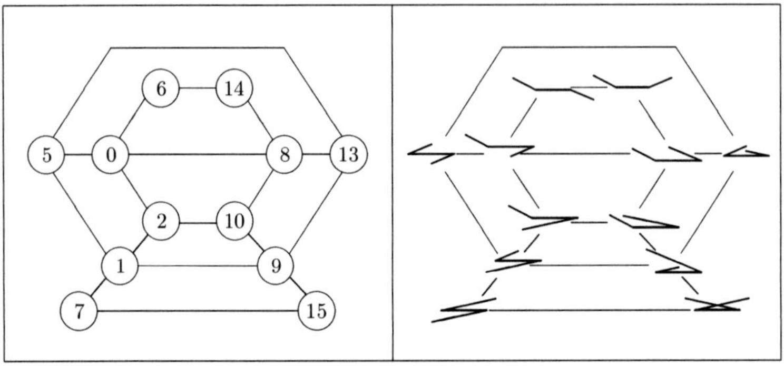

Figure 4.12: Left: the conceptual neighbourhood graph; Right: example instances; the numbers refer to the $\mathcal{TLT}_{12}$ relations

A closed polygon with $k \geq 3$ lines is described as a vector of k $\mathcal{TLT}$s, an open polygon as a vector of $k-2$ $\mathcal{TLT}$s:

$$\mathcal{TLT}_{12}(i, j, ...), i, j \in \{0, 1, 2, 5, 6, 7, 8, 9, 10, 13, 14, 15\}.$$

In order to be able to treat each arbitrary polygon with k lines, and not only those which consist of a multiple of three lines, a vector of $\mathcal{TLT}$s describes polygons in such a way that two consecutive $\mathcal{TLT}$s share two lines. Some pairs of $\mathcal{TLT}$s are not compatible with being neighbours in this way. For example, $\mathcal{TLT}_{12}(1)$ and $\mathcal{TLT}_{12}(6)$ cannot be combined, since for a combination to be possible they would have to share two adjacent lines (which is to say one angle).

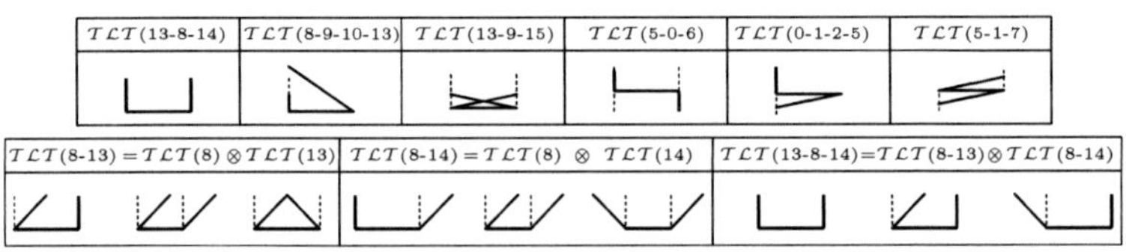

Figure 4.13: Upper row: all *singular tripartite line tracks* where both sidelines are in singular position; Lower row: the construction of $\mathcal{TLT}$(13-8-14)

But the angles of $\mathcal{TLT}_{12}(1)$ are both acute, whereas the angles of $\mathcal{TLT}_{12}(6)$ are both obtuse. A compatible combination consists of four lines, or rather of two entwined $\mathcal{TLT}$s. As for bipartite line tracks, for a given polygon, we choose from the possible cyclic permutations that description which comes first in the ordering with respect to the $\mathcal{TLT}$ numbers.

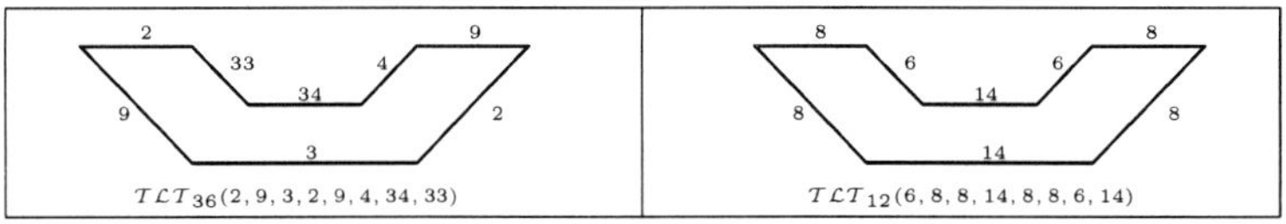

Figure 4.14: Example polygon described (anticlockwise) by tripartite line tracks

Running around the outline of a shape, it is reasonable to use $\mathcal{TLT}_{36}$ for a description. We can define the orientation as being, for example, anticlockwise regarding the two-dimensional plane, and we are able to distinguish between the inside and outside of the shape. Figure 4.14 shows an example shape with both the $\mathcal{TLT}_{12}$ and $\mathcal{TLT}_{36}$ descriptions. While $\mathcal{TLT}_{36}$ distinguishes both sides of the contour as well as between the front and back of any individual part, $\mathcal{TLT}_{12}$ distinguishes parts which are invariant with respect to rotation and reflection. From this, it follows that convex and concave parts of an outline which are otherwise similar will receive the same classification in $\mathcal{TLT}_{12}$.

$\mathcal{TLT}_6$ and $\mathcal{TLT}_{16}$

If we examine $\mathcal{TLT}_{12}$ more precisely, we notice that there are some pairs of similar relations, which differ only in the length information encoded: $\mathcal{TLT}_{12}(0)$ and $\mathcal{TLT}_{12}(2)$, for example, differ only in the length of the sideline forming an acute angle with the medial line. Taking an arbitrary instance of $\mathcal{TLT}_{12}(0)$ and lengthening the sideline which makes the acute angle, a point instant can be found where the relation changes to $\mathcal{TLT}_{12}(2)$, that is, the corresponding edge between nodes 0 and 2 of the neighbourhood graph in Figure 4.12 is visited. A similar process can be applied to other $\mathcal{TLT}_{12}$-relations, and thus we can subsume different $\mathcal{TLT}_{12}$-relations into single $\mathcal{TLT}_6$-relations as introduced in (Gottfried, 2003b):

$$\mathcal{TLT}_6(0) = [\mathcal{TLT}_{12}(0), \mathcal{TLT}_{12}(2)],$$
$$\mathcal{TLT}_6(8) = [\mathcal{TLT}_{12}(8), \mathcal{TLT}_{12}(10)],$$
$$\mathcal{TLT}_6(1) = [\mathcal{TLT}_{12}(1), \mathcal{TLT}_{12}(5), \mathcal{TLT}_{12}(7)], \text{ and}$$
$$\mathcal{TLT}_6(9) = [\mathcal{TLT}_{12}(9), \mathcal{TLT}_{12}(13), \mathcal{TLT}_{12}(15)].$$

Furthermore, it holds that $\mathcal{TLT}_6(6) = \mathcal{TLT}_{12}(6)$ and $\mathcal{TLT}_6(14) = \mathcal{TLT}_{12}(14)$.

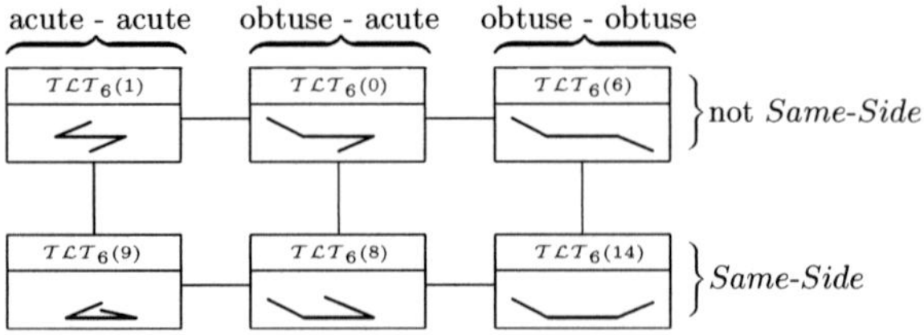

Figure 4.15: Six primitives distinguished by $\mathcal{TLT}_6$ arranged in a neighbourhood graph

The relations of $\mathcal{TLT}_6$ are depicted in Figure 4.15. Whereas $\mathcal{TLT}_{12}$ encodes some information about length as well as orientation, $\mathcal{TLT}_6$ encodes information only about orientation. It can be argued that $\mathcal{TLT}_{12}$ also encodes only orientation information, but with respect to the reference system in Figure 4.9; in a way the concept of orientation is different for $\mathcal{TLT}_{12}$ and $\mathcal{TLT}_6$. The latter is simpler since it only distinguishes acute angles, obtuse angles, and the two different sides to which a sideline can point.

In the same way that we defined $\mathcal{BLT}_4$ as the oriented version of $\mathcal{BLT}_2$, we can define the oriented version of $\mathcal{TLT}_6$ and obtain $\mathcal{TLT}_{16}$ (see Figure 4.16) which is a subset of $\mathcal{TLT}_{36}$. For $\mathcal{TLT}_{16}$, the two kinds of acute angles distinguished by $\mathcal{TLT}_{36}$ are put together into a single equivalence class.

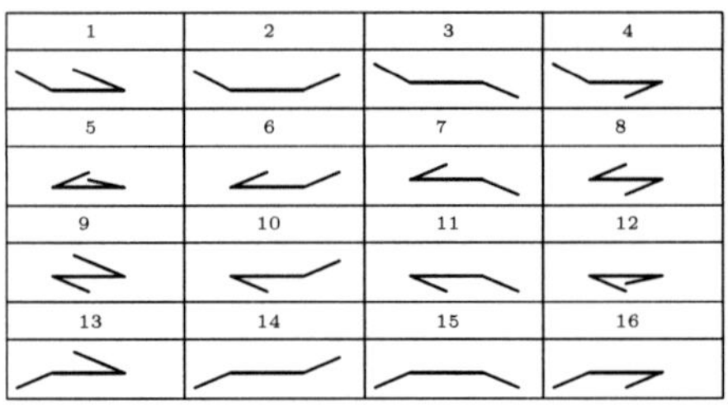

Figure 4.16: 16 oriented classes of line track arrangements with three connected lines; the medial line is oriented from left to right, the internal region is on the left of the medial line (i.e. above it, in the diagram)

4.1.3 Comparing bipartite and tripartite line tracks

Connected line tracks are divided up into oriented and non-oriented relations. The relations of $\mathcal{BLT}_4$, $\mathcal{BLT}_6$, $\mathcal{TLT}_{16}$, and $\mathcal{TLT}_{36}$ are oriented. For $\mathcal{BLT}$s the orientation is determined by one of the two lines, which is oriented towards the other; that is, considered as a vector its head is connected to the second line. Orientation must be explicitly defined for both $\mathcal{BLT}$s and $\mathcal{TLT}$s.

The relations of $\mathcal{BLT}_2$, $\mathcal{BLT}_3$, $\mathcal{TLT}_6$, and $\mathcal{TLT}_{12}$ are not oriented. $\mathcal{BLT}_3$ and $\mathcal{TLT}_{12}$ are based on the reference system in Figure 4.9. $\mathcal{TLT}_{12}$ is more expressive than $\mathcal{BLT}_3$ since it considers two adjacent angles simultaneously and also distinguishes between the two sides of a line. $\mathcal{BLT}_3$ distinguishes only obtuse angles and two kinds of acute angles. By contrast, $\mathcal{BLT}_2$ and $\mathcal{TLT}_6$ are restricted to acute and obtuse angles. $\mathcal{TLT}_6$ is more expressive than $\mathcal{BLT}_2$ for the same reasons that $\mathcal{TLT}_{12}$ is more expressive than $\mathcal{BLT}_3$. $\mathcal{BLT}_2$ is confined to two relations, i.e. the distinction between acute and obtuse angles.

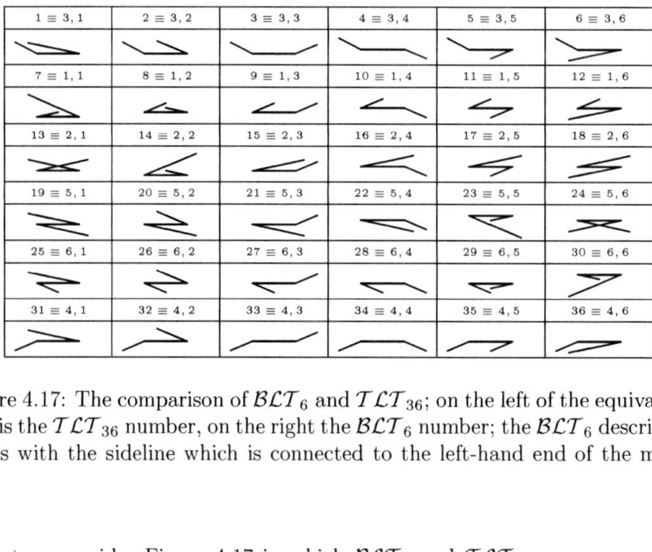

Figure 4.17: The comparison of $\mathcal{BLT}_6$ and $\mathcal{TLT}_{36}$; on the left of the equivalence sign is the $\mathcal{TLT}_{36}$ number, on the right the $\mathcal{BLT}_6$ number; the $\mathcal{BLT}_6$ description starts with the sideline which is connected to the left-hand end of the medial line

Let us consider Figure 4.17 in which $\mathcal{BLT}_6$ and $\mathcal{TLT}_{36}$ are compared — the two most expressive sets of relations. The $\mathcal{BLT}_6$ description starts with the sideline which is connected to the left of the medial line. The distinction between $\mathcal{BLT}_6$ and $\mathcal{TLT}_{36}$ arises because with $\mathcal{BLT}_6$ the reference system is initially generated from one of the two sidelines and the medial line is described relative to this sideline. With $\mathcal{TLT}_{36}$, however, this sideline is described relative to the medial line. The other sideline is described relative to the medial line in both cases. Because of this difference, some relations are not treated equally by $\mathcal{BLT}_6$ and $\mathcal{TLT}_{36}$, as demonstrated in Figure 4.18. All $\mathcal{TLT}_{36}$-relations

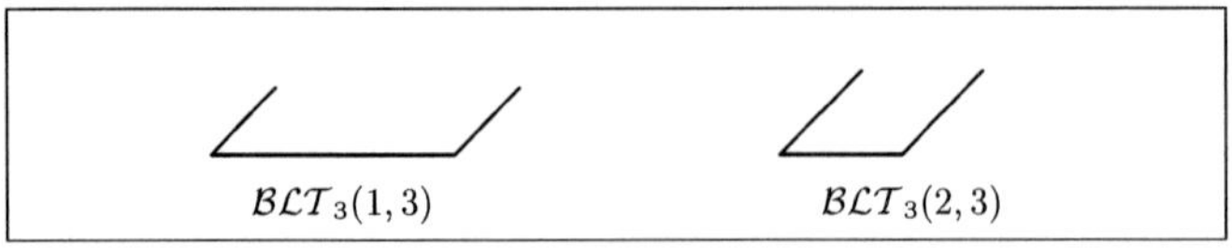

Figure 4.18: Two different polygons not distinguishable by $\mathcal{TLT}_{36}$ but which can be distinguished by $\mathcal{BLT}_3$; for both polygons the relationship $\mathcal{TLT}_{36}(9)$ holds

can also be distinguished by $\mathcal{BLT}_6$. However, some line tracks that are not distinguishable using non-oriented $\mathcal{BLT}$s can be distinguished with $\mathcal{TLT}$s.

When we compare the two least expressive, non-oriented sets of relations, namely $\mathcal{BLT}_2$ and $\mathcal{TLT}_6$, this difference is more obvious. Whereas $\mathcal{BLT}_2$ just distinguishes acute and obtuse angles, $\mathcal{TLT}_6$ makes some more sophisticated distinctions. This can be usefully demonstrated by considering polygons containing multiple $\mathcal{TLT}$s (for example, polygons with four lines), and seeing how many equivalence classes each set of representations produces. These are depicted in Figure 4.19. A more complex representation still is provided by $\mathcal{TLT}_{12}$ which distinguishes a total of 64 different equivalence classes for quadripartite line tracks. These are depicted in Figure 4.20. This also shows that some relations in $\mathcal{TLT}_{12}$ are less readily distinguishable from each other than those in $\mathcal{TLT}_6$.

To summarise, we obtain different sets of line track relations depending on the following distinctions: the usage of the orientation grid versus distinguishing acute and obtuse angles, oriented versus non-oriented primitives, and bipartite versus tripartite line tracks. Accordingly:

Definition 4.3 (Local Property)
Local polygonal properties are defined by adjacent lines in a polygon, that is, either by a line and its successor, or by a line and both its successor and its predecessor. We distinguish eight sets of relations, all of which define local properties:

- $\mathcal{BLT}_2$: *non-oriented, acute vs. obtuse*
- $\mathcal{BLT}_3$: *non-oriented, orientation grid*
- $\mathcal{BLT}_4$: *oriented, acute vs. obtuse*
- $\mathcal{BLT}_6$: *oriented, orientation grid*
- $\mathcal{TLT}_6$: *non-oriented, acute vs. obtuse*
- $\mathcal{TLT}_{12}$: *non-oriented, orientation grid*
- $\mathcal{TLT}_{16}$: *oriented, acute vs. obtuse*
- $\mathcal{TLT}_{36}$: *oriented, orientation grid*

Our definition of local properties sticks to the concept of *local feature schemes* (Meathrel & Galton, 2000). That is, by local properties we refer to relations between adjacent line segments of polygons. By contrast, (Clementini & Felice, 1997) use the term local properties to refer to those aspects of the shape which concern noise — something which should be removed by the application of regularisation techniques. In this way Clementini and De Felice do not distinguish between local aspects and global aspects of shapes, both of which could be non-accidental shape properties in sketches. It is certainly important to deal with noisy information (for example, using regularisation techniques). But like Jungert (1993) the distinction between shape representations and the processes to obtain those representations are mixed up, inevitably leading to fundamental aspects of the representation being overlooked: Jungert did not recognise relationships between shape primitives though he implicitly defines primitives comparable to $\mathcal{BLT}s$, and Clementini and De Felice fail to make a distinction between local and global shape aspects. We will now turn our attention to global properties, which are even more versatile than those aspects we have already considered in our discussion of local properties.

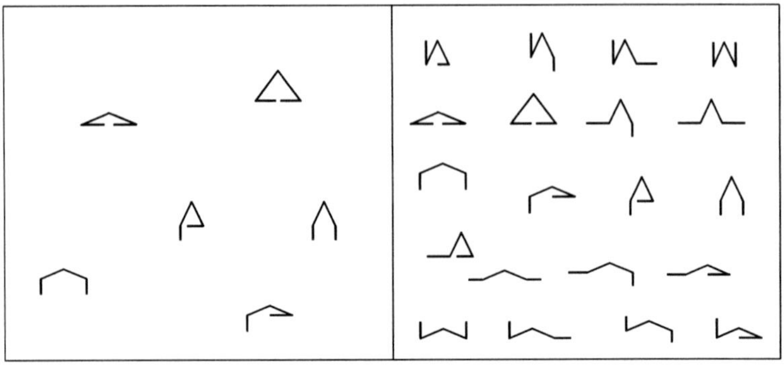

Figure 4.19: Quadripartite line tracks represented by $\mathcal{BLT}_2$ (left) which comprises only six equivalence classes, and by $\mathcal{TLT}_6$ which distinguishes 20 classes

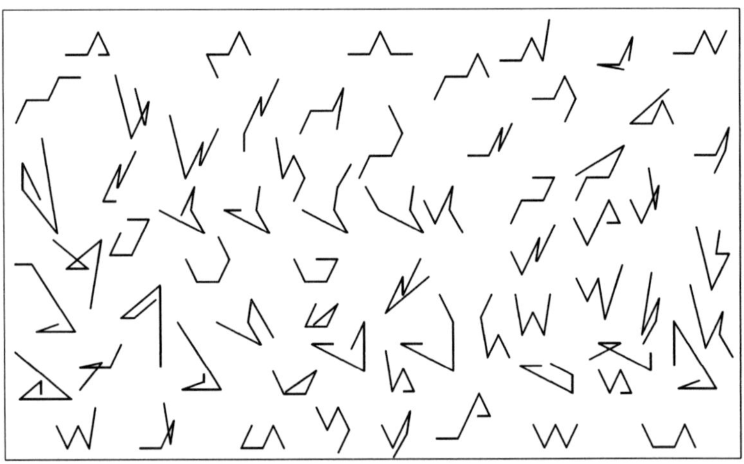

Figure 4.20: 64 quadripartite line tracks distinguished by $\mathcal{TLT}_{12}$

4.2 Global orientation

In this section, we shall introduce a *global feature scheme* which complements *local feature schemes* (Meathrel & Galton, 2000), such as the one defined in the previous section. As Figure 4.21 demonstrates, sometimes shapes look quite different even when they are based on the same $\mathcal{TLT}$s. Locally viewed, these shapes are entirely similar. Their local course is qualitatively equal and metrically similar. The longer two curves with only minor local differences are, the more such local differences will accumulate into globally significant differences. These global differences are the consequence of exploiting the $\mathcal{TLT}$s' degree of freedom in different ways.

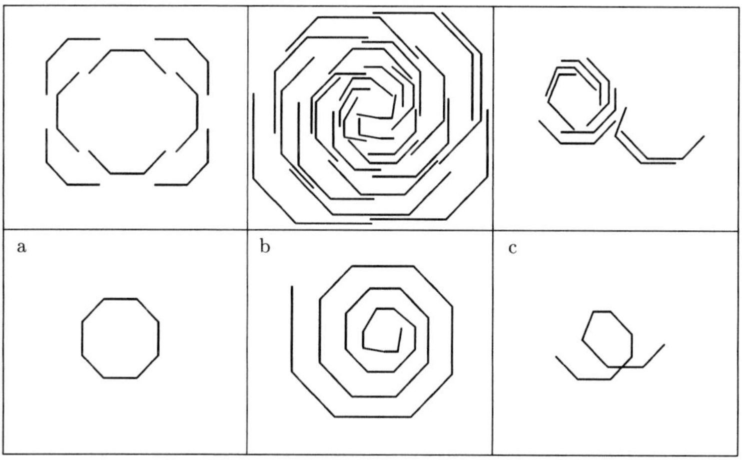

Figure 4.21: Three examples for $\mathcal{TLT}_6(14)$-contours; above: the shapes fragmented into their constitutive $\mathcal{TLT}$s

The polygons in Figure 4.21 consist entirely of $\mathcal{TLT}_6(14)$-segments. Such polygons are always bent to one side, and form patterns like circles and spirals. In order to understand these tendencies of $\mathcal{TLT}_6(14)$-polygons, we must take a closer look at $\mathcal{TLT}_6(14)$-relations. The endpoints of such line tracks lie on the same side with respect to the medial line, and both angles are obtuse. Therefore, entwined $\mathcal{TLT}_6(14)$-relations always make up a curved line which never changes its local orientation, and which can be regarded as an arc. This constancy in orientation can be comprehended if we imagine tracing a $\mathcal{TLT}_6(14)$-contour with one finger without the need for wriggling. The circle-like figures described by our finger may get larger or smaller, but they always remain approximately like a circle or an ellipse. Figure 4.21 shows three examples: depending on precise length and angle information, we obtain circle-like figures, or spirals (where the lines become consistently shorter), or loops (where the lines vary in length and angle). While circles, spirals, and loops are locally equal they obviously differ in terms of global shape properties.

Local properties correspond to local variations of the orientation which are described by line tracks. By contrast, global properties correspond to global variations of the orientation, i.e. variations concerning the relations between disconnected line segments. Therefore, global orientation variations have to be described by arrangements of disconnected intervals, i.e. by relations of $\mathcal{BA}_{23}^{8}$. Analysing which of the relations of $\mathcal{BA}_{23}^{8}$ hold, we should be able to characterise concepts such as circles, spirals, and loops.

Defining a polygonal circle (such as the one in Figure 4.21.a) as oriented anticlockwise, for two arbitrary disconnected line segments x and y it holds that

$$x_y^{\Phi} \in \{F_l, FO_l, D_l, C_l, BO_l, B_l\} \tag{4.1}$$

That is, for any polygonal circle each line segment lies to the left of any other one. This is not true of either loops or spirals, allowing us to distinguish them from circles, but not from each other. Running along a spiral from the outside inwards, lines which have already been visited can lie either to the left or to the right, but forthcoming lines are always to the left. Traversing along a loop, however, forthcoming lines can be on either side. Thus, even as simple a tool as the left-right dichotomy, combined with knowledge of the ordering of relations will allow us, when applied to nonadjacent lines, to discriminate roughly between circles, spirals, and loops. From these examples we learn that we need to know both which $\mathcal{BA}_{23}^{8}$-relations hold and how they are ordered.

Another example is shown in Figure 4.22. The left hand side of Figure 4.22 depicts a polygon which changes its local orientation at every step. This is characterised by $\mathcal{TLT}_6(1)$. But globally this polygon leads steadily in one direction — it never turns backwards. The polygon on the right hand side of Figure 4.22 is made up of $\mathcal{TLT}_6(1)$-relations, too. But this time the polygon changes its global orientation — it is in fact closed. We refer to the former polygon as *globally straight* and *locally curved*, and to the latter one again as *locally curved* but also *globally curved*. The same considerations can be applied to polygons made up of different $\mathcal{TLT}$-relations, such as the one in Figure 4.23.

What makes it difficult to analyse global properties is that a curve can be arbitrarily complex at a fine scale. A complex curve may nevertheless possess global properties which are very simple, such as a more or less straight course or an arc — for comparison, consider the case of a Mandelbrot set (Mandelbrot, 1986), where zooming in reveals ever more detailed and complex twists and turns. The right hand side of Figure 4.22 shows a good example: here we are concerned with an egg-like shape from the global point of view, whereas this polygon is locally quite complex. A reasonable way of dealing with locally complex curves is to consider them at coarser granularity levels, i.e. at

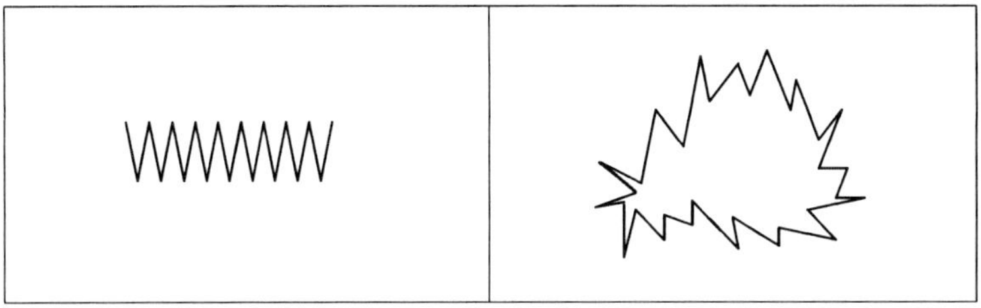

Figure 4.22: Two examples of jagged $\mathcal{TLT}_6(1)$-contours

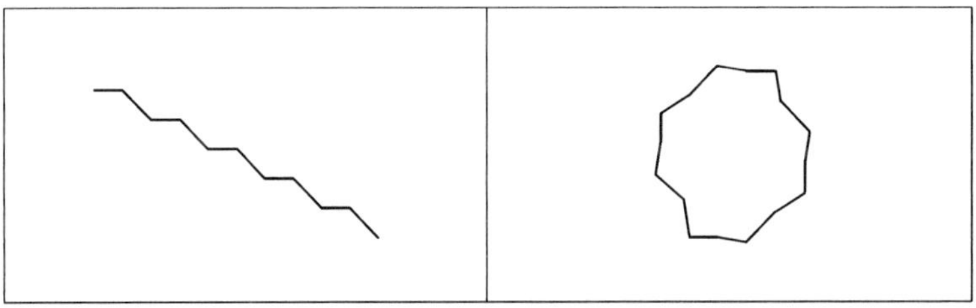

Figure 4.23: Two globally very different examples of $\mathcal{TLT}_6(6)$-contours

granularity levels where local complexities are smoothed away, for example by polygonal simplification algorithms. The analysis of global properties is easier for a smoothly curving shape than for a jagged one. However, regardless of the granularity level chosen, we are able to characterise polygons using $\mathcal{BA}$-relations. For instance, for the polygon in Figure 4.24.a it holds that $x_y = D_r^B$. By contrast, in Figure 4.24.b it holds that $x_y = F_m^F$.

It seems that *straightness* is one of the simplest global properties, but it is actually very difficult to define. We will discuss it in order to understand what difficulties arise when characterising polygons using global properties. Informally, by straight polygons we refer to polygons which lead into one direction. The simplest example is a single straight line. But how do we know that a polygon which is made up of n line segments is straight? The left hand side of Figure 4.23 suggests that one may be able to conclude this from the $\mathcal{TLT}s$ involved. But the polygon on the right hand side of Figure 4.23 demonstrates that this is not the case. Local orientation information, as encoded by $\mathcal{TLT}s$, does not allow us to derive global properties. The polygon on the left hand side in Figure 4.24 suggests that it might be possible to use $\mathcal{BA}$-relations for the purpose of recognising straightness. In this case it holds that $x_y = D_r^B$. From

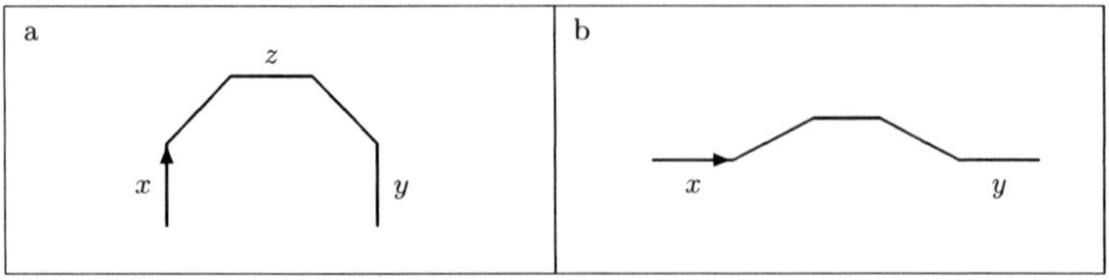

Figure 4.24: Two different polygons

this, one may come to the conclusion that there are special $\mathcal{BA}$-relations from which we can directly derive whether a polygon is generally straight (4.24.b) or otherwise (4.24.a). But as Figure 4.25 shows this is not true: In both cases it holds that $x_y = D_r^B$, but while the polygon on the left is straight from the global point of view this is clearly not true for the one on the right. Thus, we can see that it is not sufficient to analyse which $\mathcal{BA}$-relations are present in order to derive global shape properties.

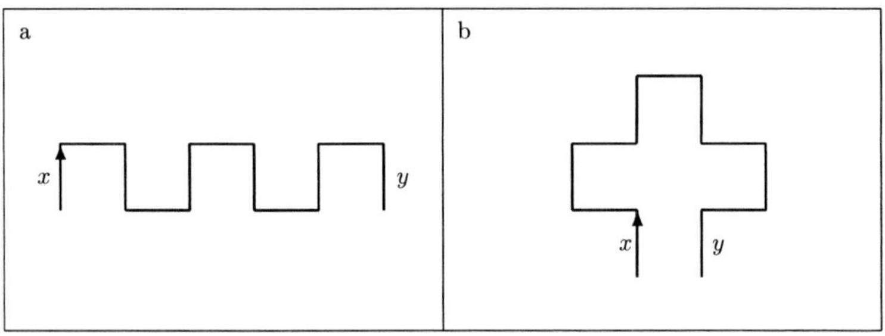

Figure 4.25: Two different polygons

The problem is that Figure 4.25 faces us with two different granularity levels. At a fine level Figure 4.25.b shows a similar pattern to Figure 4.25.a, but at a coarse level we are faced with a U-like shape. One could argue that Figure 4.25.a has an overall straight shape since it leads into one direction at a coarse granularity level. By contrast, Figure 4.25.b is not straight since it comprises this U-like shape at a coarse granularity level. This suggests that if we restrict ourselves to one granularity level, we are then able to derive straightness from $\mathcal{BA}$-relations. The question then arises whether there are ways in which we can derive global shape properties, given a polygon which comprises structures at different granularity levels.

4.2.1 $\mathcal{BA}_{23}$-courses

It is obviously necessary to consider not merely a pair of disconnected line segments but a number of such relations simultaneously, as well as their ordering within the polygon. We refer to this list of relations as the *course* with respect to line x:

$$C(x) \equiv (x^{\Phi}_{y_1}, .., x^{\Phi}_{y_n})\ x^{\Phi}_{y_i} \in \mathcal{BA}_{23}, i = 1, .., n \tag{4.2}$$

Informally speaking, the course of a line is the global orientation of the polygon from the point of view of that line. More precisely, the orientation of the polygon relative to x is defined by the positions of all lines of the polygon with respect to line x. The orientation of the whole polygon with respect to one of its lines is quite complex, and must be expressed as a list of qualitative positions — it is generally not possible to express it in terms of single positions since the global orientation comprises *a number of single positions in a special order*. Note that we define the global *orientation* using a number of *positions*.

The reference segment forms a local arrangement with those lines to which it connects at each end, this arrangement describing *local orientation* information. The local orientation of polygons has already been defined using $\mathcal{BLT}s$ and $\mathcal{TLT}s$. Therefore, lines which are adjacent to the reference line are not considered in a course, with which the *global orientation* of the polygon is to be described. We define a course as follows:

Definition 4.4 (Course)
x is line segment of a polygon. We denote its course by $C(x)$ which is defined as follows: $C(x) \equiv (x^{\Phi}_{y_1}, \ldots, \{\}, Id, \{\}, \ldots, x^{\Phi}_{y_n})$, $x^{\Phi}_{y_i} \in \mathcal{BA}_{23}; i = 1, .., n$

Note that it is not necessary to consider the empty relation explicitly in the following discussion. Definition 4.4 states, once and for all, that no $\mathcal{BA}$-relation is connected to the reference segment. For simplicity we will therefore write $C(x) = (BO_r, D_r, Id, F_r, FO_r)$ instead of $C(x) = (BO_r, D_r, \{\}, Id, \{\}, F_r, FO_r)$. This is the reason why there are only $n-1$ $\mathcal{BA}$-relations for the first and for the last course of an open polygon, and $n-2$ relations otherwise, i.e. for $1 < i < n$ in an open polygon, or any segment of a closed polygon.

Let us consider the example in Figure 4.25. For the left polygon we obtain $C(x) = (Id, D_r, BO_r, D_r, FO_r, D_r, BO_r, D_r, FO_r, D_r)$, and for the right one $C(x) = (Id, F_l, F_l, F_m, F_r, F_r, F_r, F_r, FO_r, D_r)$. The first example shows the periodicity $\overline{D_r, BO_r, D_r, FO_r}$. By contrast, the latter example shows how the course changes its orientation at relation F_m — leading first into a direction which is forward and left of x, and after F_m steadily backwards and right of x, until the relation D_r is reached.

The course of a polygon is obtained by taking all courses into consideration:

Definition 4.5 (Polygonal Course)
P is a polygon. Its course is denoted by $C(P)$ which is defined by taking simultaneously all courses of each line of P: $C(P) \equiv \bigwedge_{i=1}^{n} C(x_i)$

The course of a polygon P is simply P described in terms of all $\mathcal{BA}_{23}$-relations involved. For Figure 4.25.a we obtain the following polygonal course; each single course begins with line x:

Id	–	D_r	BO_r	D_r	FO_r	D_r	BO_r	D_r	FO_r	D_r
–	Id	–	F_r	F_r	F_m	F_r	F_r	F_r	F_m	F_r
D_r	–	Id	–	D_r	BO_l	D_l	FO_l	D_l	BO_l	D_l
B_l	B_l	–	Id	–	F_l	F_l	F_m	F_l	F_l	F_l
D_l	FO_l	D_l	–	Id	–	D_r	BO_r	D_r	FO_r	D_r
B_r	B_m	B_r	B_r	–	Id	–	F_r	F_r	F_m	F_r
D_r	BO_r	D_r	FO_r	D_r	–	Id	–	D_l	BO_l	D_l
B_l	B_l	B_l	B_m	B_l	B_l	–	Id	–	F_l	F_l
D_l	FO_l	D_l	BO_l	D_l	FO_l	D_l	–	Id	–	D_r
B_r	B_m	B_r	B_r	B_r	B_m	B_r	B_r	–	Id	–
D_r	BO_r	D_r	FO_r	D_r	BO_r	D_r	FO_r	D_r	–	Id

The empty entries, which are filled by the "–" sign, denote those positions where the empty relation is; by this means, we arrange the courses above each other so that a column always refers to the same line segment in P, i.e. one column i describes all relations of line segment i with regard to all other lines. Figure 4.25.b is described by the following polygonal course:

Id	–	F_l	F_l	F_m	F_r	F_r	F_r	F_r	FO_r	D_r
–	Id	–	D_r	BO_r	B_r	B_r	B_r	B_r	B_m	B_l
B_r	–	Id	–	F_r	F_r	F_r	FO_r	D_r	BO_r	B_r
FO_r	D_r	–	Id	–	F_l	F_l	F_m	F_r	F_r	F_r
B_m	B_l	B_l	–	Id	–	D_r	BO_r	B_r	B_r	B_r
BO_r	B_r	B_r	B_r	–	Id	–	F_r	F_r	F_r	FO_r
F_r	F_r	F_r	FO_r	D_r	–	Id	–	F_l	F_l	F_l
B_r	B_r	B_r	B_m	B_l	B_l	–	Id	–	D_r	BO_r
F_r	FO_r	D_r	BO_r	B_r	B_r	B_r	–	Id	–	F_r
F_l	F_m	F_r	F_r	F_r	F_r	FO_r	D_r	–	Id	–
D_r	BO_r	B_r	B_r	B_r	B_r	B_m	B_l	B_l	–	Id

The patterns demonstrated by the two courses are striking. The first example shows the periodicity of the polygon in the repeating pattern of positions. By contrast, in the second example the positions travel halfway around the reference line, indicating that the whole pattern makes up a kind of arc. But it must be taken into account that the explanatory power of the course of a single line is limited; taking the course of one line x, information about the course is indeterminate whenever adjacent lines have the same relation to x (for instance, the forth and fifth columns of the second row in the first course both show F_r).

Definition 4.6 (Course-indeterminacy)
x is line segment of a polygon.

(a) *Course-indeterminacy arises if* $\exists_{y \neq x} : x_y = x_{y'}$.

(b) *The number of relations which are equal and adjacent determines the degree of indeterminacy.*

Course-indeterminacy in $C(x)$ can be compensated for by those lines of the same polygon which define courses which are not indeterminate at the same positions. This can be derived from the course of the polygon, i.e. by looking for a line y which has differing relations in its course $C(y)$ where $C(x)$ is indeterminate.

In the rest of this chapter we will investigate what kinds of global property can be derived from a polygon's course.

Changes in direction

In this section possible courses are systematically investigated. For this purpose, we take the reference system of Figure 3.8 and annotate it with the $\mathcal{BA}_{23}$-relations, as shown in Figure 4.26. Note that we can use a subset $\mathcal{BA}_{13} \subset \mathcal{BA}_{23}$ which comprises only 13 relations, because *contains*–relations, i.e. the relations $FC_l, C_l, BC_l, FC_r, C_r, BC_r$, take a course which can be described using a series of different $\mathcal{BA}_{13}$-relations; for instance, in the configuration space $\mathcal{BA}_{13}$ the course C_l runs along the relations $(B_l, BO_l, D_l, FO_l, F_l)$. While it is true that C_l can be described by (BO_l, D_l, FO_l), if we are interested in those relations of $\mathcal{BA}_{13}$ which are somehow related to a *contains*-relation like C_l (in order to determine conceptual neighbours, for instance) it is necessary to include B_l and F_l since B_m and F_m are conceptual neighbours of C_l and of B_l and F_l but not of BO_l and FO_l; therefore, (BO_l, D_l, FO_l) does not describe the course of C_l appropriately. The same considerations apply to *overlap–medial*–relations, such as $FO_{ml}, FO_{mr}, BO_{ml}, BO_{mr}$. These ten relations are used only where they are necessary to specify specific global properties. Figure 4.27 shows the neighbourhood graph which describes possible neighbours when considering adjacent line segments in a polygon, i.e. $\mathcal{BA}$-relations which can be adjacent in a course.

We will now consider courses which have no change in direction, which change once, and which change twice or more often, in that order, calling our example polygon P. From the point of view of a line $x \in P$, single directions refer to those cases where all other lines are at the same position with respect to x, i.e. for x there exists no change in direction along P. We refer to this case as a course of grade zero, as indicated by the index:[2]

[2] $\forall_{y \in P}$ implies that we exclude the direct predecessor and direct successor of the reference segment $x = x_i$ — this is always assumed in the following discussion; otherwise we would always have to state additionally $y \neq x_{i-1} \wedge y \neq x_{i+1} \wedge z \neq x_{i-1} \wedge z \neq x_{i+1}$.

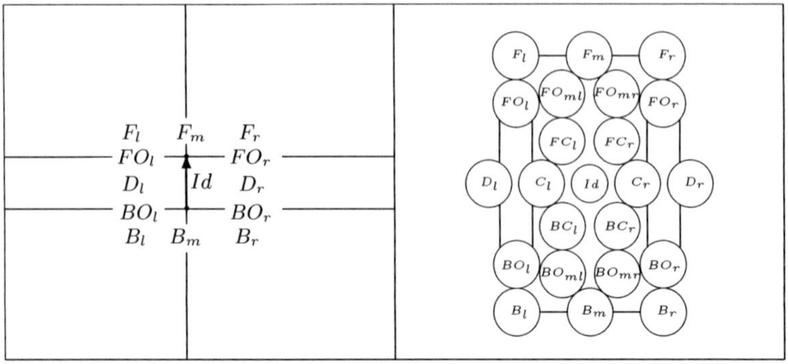

Figure 4.26: A line segment defines different qualitative positions which form the configuration space of courses; for comparison, on the right hand side is the neighbourhood graph which shows all $\mathcal{BA}_{23}$-relations

$$C_0(x) \equiv \forall_{y \in P: y \neq x} \forall_{z \in P: z \neq x} : x_y = x_z \tag{4.3}$$

Provided that it is clear that all lines involved are part of the same polygon P we can simplify the formula as follows:

$$C_0(x) \equiv \forall_{y \neq x, z \neq x} : x_y = x_z \tag{4.4}$$

Considering courses of grade zero, there are five different directions to be distinguished, all other directions being symmetric to these five cases: F_m, F_r, FO_r, D_r, and BO_r. Figure 4.28 depicts examples of these. In order to concisely characterise a course we simplify these descriptions and write $C(x) = F_m$ instead of $C(x) = (F_m, F_m, F_m, F_m, F_m, F_m)$, i.e. adjacent relations which are equal can be deleted in order to describe the overall course regarding a reference line x.

$C_0(x)$ can be true for a section of P which starts at line segment v and ends at line segment w. Accordingly, we can define:

$$C_0(x, v, w) \equiv \forall_{v \leq y \leq w, v \leq z \leq w} : x_y = x_z \tag{4.5}$$

In fact, by $C_0(x)$ we mean $C_0(x, x'', x_n)$, with x'' denoting the successor of the successor of x, and x_n the last line segment of P.

It is of course possible for the polygon to be quite complex in the section where all lines are equally placed with respect to x. $C_0(x)$ states only that from the point of view of x nothing interesting happens, this viewpoint being defined by the reference system which is induced by x. That is, $C_0(x) = D_r$ includes a case like that depicted on the left hand side of Figure 4.29, as well as almost straight polygons similar to the right hand side. On the other hand,

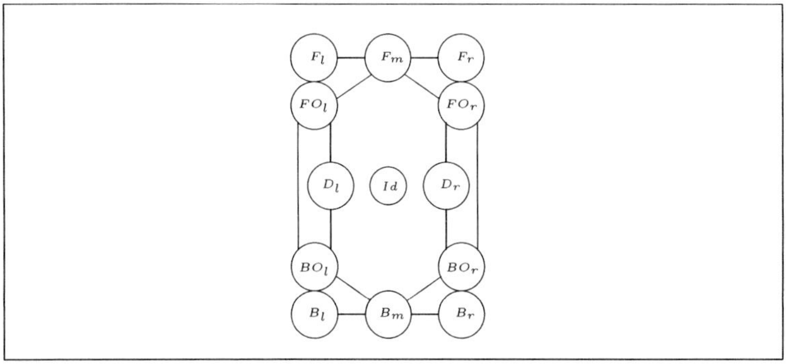

Figure 4.27: The conceptual neighbourhood graph of $\mathcal{BA}_{13}$

this example shows that $C_0 = D_r$ appropriately represents those cases where there is shape information only to the right of x, level with x, and nowhere else — which may be exactly what we are interested in. In order to distinguish the polygons in Figure 4.29 the course of at least one additional line segment must be considered. That is, it is crucial to know how many courses (and from which lines) must be taken into account in order to compensate for course-indeterminacy. In general, this question can be answered only by analysing the polygon's course, i.e. every single course of P.

Courses which change their direction once are defined as follows:

$$C_1(x) \equiv \exists_{w \neq x} : C_0(x, x'', w) \wedge C_0(x, w', x_n) \wedge x_w \neq x_{w'} \tag{4.6}$$

A section for which $C_1(x, v, w)$ holds can thus be defined in terms of equation 4.5. Courses which change twice are defined as follows:

$$C_2(x) \equiv \exists_{w \neq x} : C_0(x, x'', w) \wedge C_1(x, w', x_n) \wedge x_w \neq x_{w'} \tag{4.7}$$

It is actually unimportant whether C_1 follows C_0 or the other way round; it is only necessary that C_2 contains two changes in direction — a section with one change, C_1, a section without any change, C_0, and a change between C_1 and C_0. Generally, for k changes in direction we can define recursively:

$$C_k(x) \equiv \exists_{w \neq x} : C_0(x, x'', w) \wedge C_{k-1}(x, w', x_n) \wedge x_w \neq x_{w'}, k \geq 1 \tag{4.8}$$

A more general definition cuts the polygon at an arbitrary line segment l:

$$C_k(x) \equiv \exists_{w \neq x} : C_l(x, x'', w) \wedge C_{k-l-1}(x, w', x_n) \wedge x_w \neq x_{w'}, k \geq 1, l < k \tag{4.9}$$

Up to now, we have acted on the assumption that we are interested in only one direction concerning the course of x. The complete polygon can be treated

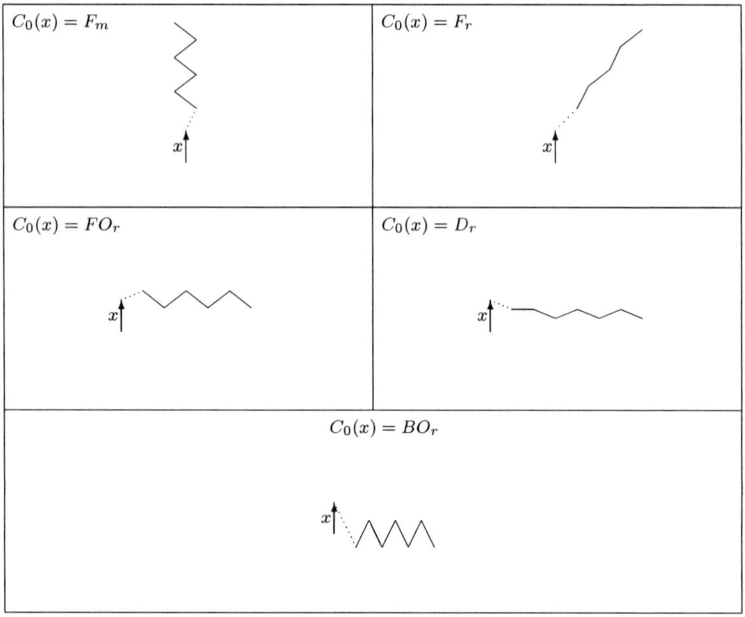

Figure 4.28: Five different polygons describing five different courses of grade 0 with respect to line segment x

under this assumption only if it holds that $x = x_1$. Otherwise, by our definition of C_k we have ignored any section of the polygon before line segment x. In order to consider both the section before line segment x and the section after it we write $C_k(x_1, x_{i-2}, x_i, x_{i+2}, x_n)$, with reference line $x = x_i$. We must then distinguish whether the line segments before and after x_i are inverse (see chapter 3.2.2) to each other or not:[3]

$$\begin{aligned} &C_k(x_1, x_{i-2}, x_i, x_{i+2}, x_n) \equiv \\ &\quad (x_{i-2} = x_{i+2}^{-1} \wedge C_l(x_i, x_1, x_{i-2}) \wedge C_0(x_i, x_{i-2}, x_{i+2}) \wedge C_{k-l}(x_i, x_{i+2}, x_n)) \vee \\ &\quad (x_{i-2} \neq x_{i+2}^{-1} \wedge C_l(x_i, x_1, x_{i-2}) \wedge C_1(x_i, x_{i-2}, x_{i+2}) \wedge C_{k-l-1}(x_i, x_{i+2}, x_n)) \\ &\qquad \text{with } 1 < i < n; k \geq 0; (0 \leq l < k) \vee (l = k = 0) \end{aligned} \tag{4.10}$$

[3]These actually concerns the predecessor of the predecessor of x_i and the successor of the successor of x_i, since the course comprises empty relations for the immediately adjacent lines; see Definition 4.4.

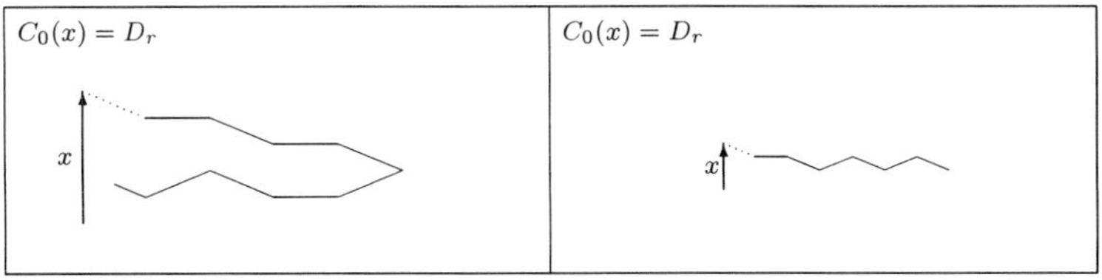

Figure 4.29: Two different polygons describing the same course with respect to line segment x

In the first case the transition via the reference segment, $C_0(x_i, x_{i-2}, x_{i+2})$, does not count as a change in direction, but in the latter case, $C_1(x_i, x_{i-2}, x_{i+2})$, it does. This is because if the segment x_{i-2} before x_i is inverse to the segment x_{i+2} after x_i there is no change in direction. A change occurs when these two segments are not inverse relative to each other. Note that for closed polygons this formula always applies since x_i always has predecessors and successors when the polygon is closed regardless of what i is.

Where we are concerned with the whole polygon we can write simply $C_k(x)$ which is a convenient abbreviation for $C_k(x_1, x_{i-2}, x_i, x_{i+2}, x_n)$. The position of the identity relation shows where the reference segment x is. Together with Definition 4.4 this allows us to say that

$$C_k(x) = (x^{\Phi}_{y_1}, \ldots, x^{\Phi}_{y_n}); x^{\Phi}_{y_i} \in \mathcal{BA}_{23} \cup \{\}; i = 1, \ldots, n; k \geq 0 \qquad (4.11)$$

Equal neighbouring relations can be replaced with a single relation if we are interested in a more concise description; we then obtain:

$$C_k(x) = (x^{\Phi}_{y_1}, \ldots, x^{\Phi}_{y_m}); x^{\Phi}_{y_i} \in \mathcal{BA}_{23} \cup \{\}; i = 1, \ldots, m; 1 \leq m \leq n; k \geq 0 \quad (4.12)$$

Courses have been characterised by their changes in direction, the number of such changes determining the grade of a course. These definitions can be summarised:

Definition 4.7 (Changes in Direction)
Given polygon P and course $C(x)$ with $x = x_i$. k changes are defined by:

$k = 0 \wedge (i = 1 \vee i = n) \wedge open(P)$:
$C_k(x) \equiv \forall_{y \neq x, z \neq x} : x_y = x_z$

$k \geq 1 \wedge i = 1 \wedge 3 < j < n \wedge open(P)$:
$C_k(x) \equiv C_l(x_i, x_3, x_j) \wedge C_{k-l-1}(x_i, x_{j+1}, x_n) \wedge x_{x_j} \neq x_{x_{j+1}}$

$k \geq 1 \wedge i = n \wedge 1 < j < n-2 \wedge open(P):$
$C_k(x) \equiv C_l(x_i, x_1, x_j) \wedge C_{k-l-1}(x_i, x_{j+1}, x_{n-2}) \wedge x_{x_j} \neq x_{x_{j+1}}$

$k \geq 0 \wedge (((1 < i < n) \wedge open(P)) \vee ((1 \leq i \leq n) \wedge closed(P))):$
$$C_k(x) \equiv (x_{i-2} = x_{i+2}^{-1} \wedge C_l(x_i, x_1, x_{i-2}) \wedge C_0(x_i, x_{i-2}, x_{i+2}) \wedge C_{k-l}(x_i, x_{i+2}, x_n)) \vee$$
$$(x_{i-2} \neq x_{i+2}^{-1} \wedge C_l(x_i, x_1, x_{i-2}) \wedge C_1(x_i, x_{i-2}, x_{i+2}) \wedge C_{k-l-1}(x_i, x_{i+2}, x_n))$$

with $(0 \leq l < k) \vee (l = k = 0)$

Finally, two observations should be made:

(i) Given any course, $C(x)$, all other courses of the same polygon are constrained in that the relation of their reference segment to x is already known and the converse operation, $\breve{x}_y = y_x$, can be used to determine the position of x with respect to all other line segments.

(ii) Where two courses, $C(x)$ and $C(y)$, of the same polygon are incompletely given, these courses can be completed by composition: if, for example, $C(x)$ lacks the relation to a line segment z but $C(y)$ includes y_z, we can obtain x_z by $x_z = x_y \circ y_z$.

Multitude of courses

In order to get an idea of the expressiveness of courses we shall count the number of different courses which exist for each grade. Using $\mathcal{BA}_{12}$, there are 24 possible C_0 courses, 12 where the reference line is x_1 and 12 where it is x_n. As soon as the reference segment x is neither the first line of the polygon nor the last one there must be at least one change of direction unless the lines x_{i-2} and x_{i+2} which enclose the reference segment are inverse to each other (see Definition 4.7).

Where there is a change in direction the local relation controls how many successive relations exist. This can be seen in the neighbourhood graph in Figure 4.27: for those relations which are wholly contained in one of the six regions, i.e. for F_l, F_r, D_l, D_r, B_l, and B_r, there are two possible successors, and for the other six relations there are four possible successors. There are thus 3 possible successors on average.

In the case of C_1 we obtain the following courses. For $x = x_1$ or $x = x_n$ there must be a change in direction somewhere between the two, with an average of three possible directions for any change. As there are 12 possibilities if one starts with reference segment x_1 there are $12 * 3 = 36$ courses. The same holds for reference segment x_n, making a total of $2 * 36 = 72$ possibilities. When the reference segment x is somewhere in between, there are $12 * 12 = 144$ possible changes, less those cases for which it holds that $x_{i-2} = x_{i+2}^{-1}$, (of which there are 12, since every relation of $\mathcal{BA}_{12}$ has an inverse), making $144 - 12 = 132$ courses; the change in direction with Id in between, i.e. $..., x_{i-2}, x_i, x_{i+2}, ...$ is regarded as one single change (or as no change if $x_{i-2} = x_{i+2}^{-1}$). From this it follows that there are $132 + 72 = 204$ different courses of grade 1.

How many courses exist for C_k, where $k \geq 1$? We have already seen that there are twelve possibilities when starting or finishing with the reference segment x. As there are, on average, 3 kinds of changes possible, there are $12 * 3^k$ courses $C_k(x)$ for each of $x = x_1$ and $x = x_n$, i.e. $2 * (12 * 3^k)$ altogether. For $x_1 < x < x_n$, there are 132 possibilities for a change at the position of x itself and $k - 1$ further changes in direction (each one of 3 possible kinds) before and after x: $132 * 3^{k-1}$. Altogether there are $2 * (12 * 3^k) + (132 * 3^{k-1})$ possible courses with k changes. The number of courses increases exponentially with the number of changes involved, and we therefore conclude that the concept of courses is an expressive means of distinguishing a wide range of polygons, covering a multitude of variations.

For $k = 2$ there are 612 courses; their are 1836 for C_3 and 5508 courses for C_4. If we include the *contains*-relations and *overlap*-relations then there are even more possibilities. As different relations have a different number of neighbours we then have to determine a new average for possible changes, in order to obtain the number of possible courses for any C_k. Furthermore, we have considered only the number of courses with k changes. We could additionally take into account all combinations of courses which form different polygonal courses. For this purpose, the consistency of the courses involved in one polygonal course can be maintained by the algebraic properties of $\mathcal{BA}$ (see chapter 3.2.2 and (Gottfried, 2004a)).

It is interesting to note that, whereas polygons are frequently analysed and compared in terms of their number of lines involved, our definition of C_k is actually independent of the number of lines. This is important since most global properties, such as straight, angled, round etc., are also independent of the number of lines involved.

Circulation, scope, extent

$\mathcal{BA}_{23}$-courses have been introduced, as has a simple classification of courses consisting of determining how often they change their directions. Such changes have been defined with regard to $\mathcal{BA}_{23}$, i.e. a change in direction corresponds to a change of a $\mathcal{BA}_{23}$-relation into another one for two adjacent lines with respect to the same reference line. Considering only the number of changes, the conclusions which can be derived about a course are quite limited. It would be more useful to take into account where these changes occur and where the course runs with respect to the reference line. Such positions are given by the $\mathcal{BA}_{23}$-relations, but what does a list of twenty (or even many more) relations tell us? It is obviously necessary to reduce the information load in order to obtain concise characterisations of polygons.

Each course runs in one or more regions around the reference line and a simple description of a course consists of determining the area where the course is to be found. If the course is continuous, there are a number of neighbouring

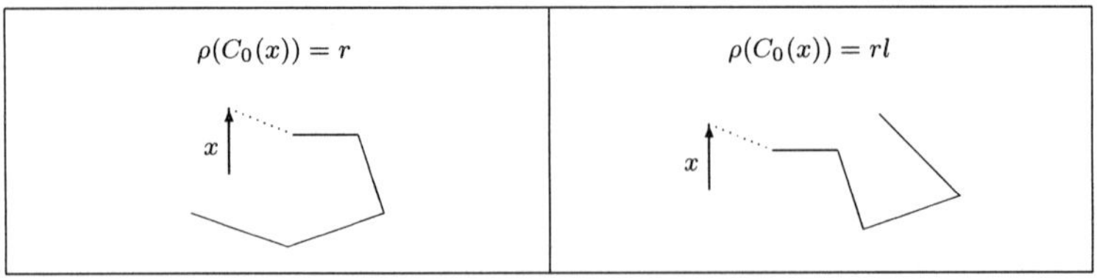

Figure 4.30: Circulation directions around a reference line

relations which connect the beginning of the course with its ending, covering a single connected area. This area can be described by two $\mathcal{BA}_{12}$-relations; the one which marks the start of the area and that which marks its end. The circular arrangements of $\mathcal{BA}$-relations allow us to circulate either clockwise or anticlockwise around the reference line. Therefore, to follow the area of the course (and not the area outside it), it is necessary to give the circulation-direction (either left of the reference line or right of it). This is defined as follows:

Definition 4.8 (Circulation)
x is line segment of a polygon P. The circulation between two neighbouring lines of P, y and y', is left of x, in short $\rho(x_y, x_{y'}) = l$, if the path from the first relation x_y to the second relation $x_{y'}$ runs anticlockwise around x; otherwise, it is right of x, in short r. If the direction does not change it holds that $\rho(x_y, x_{y'}) = \varepsilon$.

Figure 4.30 shows two examples. On the left hand side the course of x circulates entirely right of x. On the right hand side it also runs right of x but it then turns back and runs left of x. For the circulation along a section with $m > 1$ lines it holds that

$$\rho(C(x, y, z)) \in \{l, r, \varepsilon\}^{m-1} \tag{4.13}$$

For the circulation of the entire course there are three cases to be distinguished depending on whether the entire course is open or closed (and, if open, continuous or interrupted by the reference segment):

1. (a) $x = x_1 \wedge open(P)$:
   $\rho(C(x)) = \rho(C(x, x_3, x_n)) \in \{l, r, \varepsilon\}^{n-3}$

   (b) $x = x_n \wedge open(P)$:
   $\rho(C(x)) = \rho(C(x, x_1, x_{n-2})) \in \{l, r, \varepsilon\}^{n-3}$
2. $x_1 < x < x_n \wedge open(P)$:
   $\rho(C(x)) = \rho(C(x, x_1, x_{i-2}))\rho(C(x, x_{i+2}, x_n)) \in \{l, r, \varepsilon\}^{n-4}$ with $x_i = x$

3. $x_1 \leq x \leq x_n \wedge closed(P)$:
 $\rho(C(x)) = \rho(C(x, x_1, x_{i-2}))\rho(C(x, x_{i+2}, x_n)) \in \{l, r, \varepsilon\}^{n-3}$ with $x_i = x$

For each $C_0(x)$ it holds that $\rho(C_0(x)) = \varepsilon$ since there is no change of circulation direction. For each $C_k(x)$ with $k > 0$ there is at least one direction to be mentioned. For example, $\rho(F_l FO_l) = l$, $\rho(FO_l F_l) = r$, $\rho(F_l F_m) = r$, $\rho(F_l FO_{mr} BO_r) = rr$, $\rho(F_l FO_{mr} FO_r FO_r D_r) = rlr$, and $\rho(FO_l F_l IdD_r BO_r) = rr$. Equal neighbouring directions can be omitted in order to obtain only the changes, i.e. changes between left and right, or anticlockwise and clockwise, respectively. It then holds that the number of changes between left and right of $\rho(C_k(x))$ is less than or equal to k.

We refer to the range through which the course runs as the scope of that course:

Definition 4.9 (Scope)
x is a line segment of a polygon and $C(x)$ is its course. The entire range of relations where $C(x)$ runs along is called the scope of the course, in short $\sigma(C(x))$. For $r_1, r_2, r_3, r_4 \in \mathcal{BA}_{12}$, and $\rho, \rho_1, \rho_2 \in \{l, r, \varepsilon\}$ the scope is defined by

$$\sigma(C(x)) \equiv \begin{cases}]r_1, \rho, r_2], & x = x_1 \\ [r_1, \rho_1, r_2[\ ]r_3, \rho_2, r_4], & x = x_i,\ i = 2, .., n-1 \\ [r_1, \rho, r_2[, & x = x_n \end{cases}$$

Note that, taking only r_1 and r_2, the circulation-direction would be left undecided, since the reference system used is circular. The brackets indicate where the reference segment is, the scope being open around it. Thus, $]r_1, \rho, r_2[$ describes the scope of a closed polygon since the reference segment is touched by both ends of the scope. The scope can also be determined for sections of the course:

$$\sigma(C(x, y, z)) \equiv [r_1, \rho, r_2], r_1, r_2 \in \mathcal{BA}_{12}, \rho \in \{l, r, \varepsilon\} \tag{4.14}$$

For the polygon on the left in Figure 4.30 it holds that $\sigma(C(x)) =]D_r, r, B_m]$, and for the one on the right it holds that $\sigma(C(x)) =]FO_r, r, B_r] =]B_r, l, FO_r]$. The latter shows that the start-relation r_1 and the end-relation r_2 of the scope do not necessarily coincide with the first relation and the last relation in the course.

The scope is of interest whenever it is sufficient to give the position of the whole polygon with respect to a reference line without explicitly describing the shape of the course. Note that the start and end are each given by one relation of $\mathcal{BA}_{12} (= \mathcal{BA}_{13} \setminus \{Id\})$ but that the scope may include every relation of $\mathcal{BA}_{23}$. For instance, the two courses FO_{mr} and $F_l F_m F_r FO_r$ have the same scope, namely $]F_l, r, D_r]$. In Figure 4.31 we give all possible scopes starting at position F_l relative to a reference line x and circulating right of x. Each scope describes which relations of $\mathcal{BA}_{23}$ are realisable with respect to the given reference line.

Attention should be paid to the distinction between $]F_l, \varepsilon, F_l]$ and $]F_l, r, F_l]$: in the first case the course runs only somewhere along F_l, i.e. there is no need to give a circulation-direction (l or r) since the course does not change

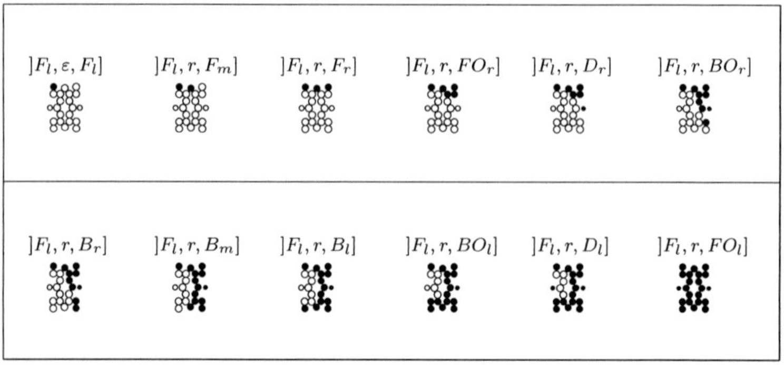

Figure 4.31: The scope of courses regarding a reference line, all starting by F_l having the direction r; the black circles indicate which relations can hold in the respective scope

its circulation-direction regarding the reference line; in the latter case it runs completely around the reference line, and it holds that $]F_l, l, F_l] =]F_l, r, F_l]$, or generally:

Dependency 4.1 $\forall_{x_y \in \mathcal{BA}_{12}} :]x_y, l, x_y] =]x_y, r, x_y]$.

Proof: each scope $]x_y, \rho, x_y]$ with $x_y \in \mathcal{BA}_{12}$ and $\rho \in \{l, r\}$ includes all relations of $\mathcal{BA}_{12}$ since all relations of $\mathcal{BA}_{12}$ are to be visited when running along exactly one direction from any relation x_y around x until returning to the same relation x_y, regardless of whether running left or right around the reference line.

$\square$

The scopes which are described in Dependency 4.1 are also referred to as the universal scopes since all relations of $\mathcal{BA}_{23} \setminus Id$ are realisable with these scopes. As shown in Figure 4.31, the scope $]F_l, r, FO_l]$ is universal. Universal scopes have the largest possible extent which refers to the number of different relations passed through by the course:

Definition 4.10 (Extent)
$\sigma(C(x))$ is the scope of course $C(x)$ of Polygon P. The distance between the start-relation and the end-relation of the scope $\sigma(C(x))$ is called the extent of the scope, or simply the extent of course $C(x)$, denoted by $\eta(C(x))$. The extent is determined with the relations of $\mathcal{BA}_{12}$, and it holds that

$$\eta(C(x)) \in \{1, 2, .., 12\}$$

The extent of the polygon is the average extent of all courses:

$$\eta(P) = \frac{\sum_{i=1}^{n} \eta(C(x_i))}{n}$$

The neighbourhood graph of $\mathcal{BA}_{12}$-relations, which determines the distance between the start-relation and the end-relation, is shown on the left hand side of Figure 4.26. For example, the extent of the scope $]F_l, \varepsilon, F_l]$ is $\eta(F_l) = 1$, and for $]F_l, r, BO_r]$ we can see that $\eta(F_l FO_{mr} BO_r) = 6$ (where the extent is obtained by taking the following relations into account: F_l, F_m, F_r, FO_r, D_r, and BO_r). For both polygons in Figure 4.30 it holds that $\eta(C(x)) = 5$. Clearly, the extent of the universal scope is 12.

Finally, it is worth noticing that the scope and extent of a course taken together form a concise summary of that course; a much greater simplification than merely putting together equal relations as mentioned earlier.

Coarse relations

We are looking for a way of reducing the information content in order to arrive at a concise description. This can be achieved by putting together different $\mathcal{BA}_{23}$-relations which make up the same direction at a coarser granularity level. Such coarser directions form subsets of $\mathcal{BA}_{23}$. Leaving out the identity relation, there are $2^{22} = 4194304$ possible subsets. Some subsets are particularly useful as coarse directions; for example, a course may run somewhere left of the reference line:

$$C_k(x) = l \equiv \forall_{y \neq x} : x_y \in \{F_l, FO_l, D_l, C_l, BO_l, B_l\} \tag{4.15}$$

Figure 4.32 shows those six relations which are completely left of the reference line. They combine to $2^6 = 64$ subsets of $\mathcal{BA}_{23}$ which form the basis of several different courses of different grades, all of them running left of the reference segment — though the restriction to connected courses in which only neighbouring relations can follow each other reduces this number.

$C_k(x) = r$, $C_k(x) = F$, $C_k(x) = B$, or other combinations of $\mathcal{BA}_{23}$-relations can be defined in the same way; especially, *front-middle*[4]:

$$C_k(x) = Fm \equiv \forall_{y \neq x} : x_y \in \{F_m, FO_{ml}, FO_{mr}, FC_l, FC_r\} \tag{4.16}$$

and *back-middle* in the same way; *middle* is then defined by:

$$C_k(x) = m \equiv Fm \vee Bm \tag{4.17}$$

A course can then be described at a less fine granularity level:

$$C(x) = (Fm, r, Bm, l, Fm) \tag{4.18}$$

[4]Note that the course Fm which we define at the present is to be distinguished from the $\mathcal{BA}_{23}$-relation F_m. F_m may be part of a course Fm, but does not have to be.

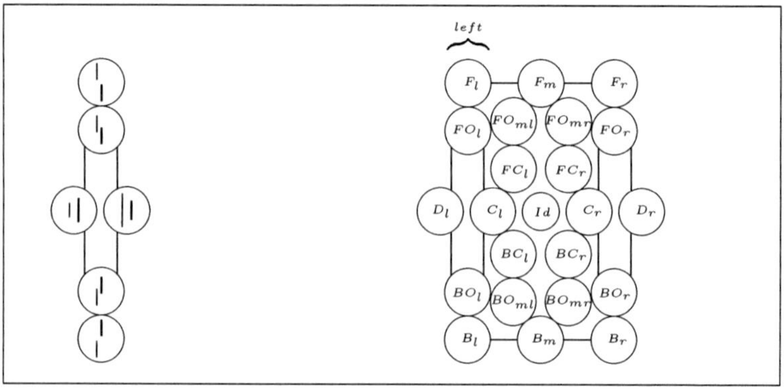

Figure 4.32: A subset of the neighbourhood graph of $\mathcal{BA}_{23}$ which contains only relations which are left of the reference line

and

$$\eta(Fm, r, Bm, l, Fm) = 12 \tag{4.19}$$

By this means, complex concepts can be described by fewer relations than when taking individual $\mathcal{BA}_{23}$-relations. The previous example (4.18) shows how a course can be represented with the primary intervals enclosing the reference line. Note that this description is more abstract and less specific than $\mathcal{BA}$ since Fm stands for a number of different relations, such as F_m or FO_{mr}. But it must be pointed out that each such concise description implies that all relations involved, coarse or fine, are conceptual neighbours to adjacent relations, since it is assumed that a continuous curve without any gaps is represented. This is the reason why there are actually fewer than 2^{22} possible subsets describing coarse directions.

Having described one curve by atomic $\mathcal{BA}$-relations and another one by coarse relations, these curves can be compared particularly well using their scopes. As for atomic $\mathcal{BA}$-relations, the scope of courses which are described by coarse relations are characterised by $\mathcal{BA}_{12}$-relations.

4.2.2 Global properties

In the last section general characteristics of courses have been discussed. We will now move on to consider properties of polygons, such as whether a polygon is convex, a line segment is a global extremum, and the curvature of polygons. In addition to these general properties, there are also a number of more specific properties which can be defined on the basis of specific $\mathcal{BA}_{23}$-relations, such as left-wing and right-wing oriented bends as well as the straightness of polygons.

General properties

There are a number of properties which can be defined without referring to any specific $\mathcal{BA}_{23}$-relations involved. In this sense, we refer to such properties as to general properties.

Curvature
Having introduced the course of a polygon one important property can directly be derived, namely the polygon's curvature:

Definition 4.11 (Curvature)
x is a line segment of a polygon P and $C(x)$ is its course.

(a) *The curvature of $C(x)$, in short $\xi(C(x))$, denotes the number of changes of $\mathcal{BA}_{23}$ concerning adjacent relations in $C(x)$. For $C_k(x)$ it holds that $\xi(C(x)) = k$.*

(b) *The curvature of P is defined as the average of the curvatures of all courses, i.e. by $\xi(P) = \sum_{i=1}^{n} \frac{\xi(C(x_i))}{n}$.*

Dependency 4.2
$\xi(C(x)) < n - 2$

Proof: as the curvature is defined by the number of changes of $\mathcal{BA}_{23}$ relations of adjacent line segments, ξ is bounded by the maximum number of possible changes involved, i.e. by the number of lines involved minus 3 for open polygons, and $n - 4$ for closed polygons.

□

By the term curvature we mean a qualitative concept of curvature, such as the one defined in (Gottfried, 2003b). There, it is argued that $\mathcal{TLT}s$ define a stylised concept of curvature and that curvature information is represented by two adjacent curvature extrema, i.e. two successive angles on the polygonal contour. Instead of using single metrical curvature-values at each contour point, we are then concerned with *extensive curvature information*, being expanded over more or less long contour segments. This becomes particularly useful when dealing with sketches where small, metrical curvature-values often reflect accidental properties, whereas necessary properties in sketches correspond more closely to qualitative curvature-relations. Examined in terms of $\mathcal{BA}$-relations the concept of qualitative curvature information is the same as for $\mathcal{TLT}s$.

Extremum
A global extremum is a line segment x with its course being completely on the left, on the right, in the front, or in the back. Before considering this, we have to discuss the convexity of polygons. For this purpose, we assume an anticlockwise orientation of a simple, closed polygon P.

Definition 4.12 (Convexity)
P is a polygon. A closed and simple polygon is convex, for short $\mathcal{C}(P)$, if for two arbitrary points contained in the interior of the polygon it holds that the straight line which connects these points is wholly contained in the polygon. Otherwise the polygon is concave.

It then holds

Dependency 4.3
$\mathcal{C}(P) \Leftrightarrow (\forall_{x \in P} : \forall_{y \in P} : y \neq x \Leftrightarrow x_y = l)$

Proof: reduction on the triangle orientation: for three adjacent points of a convex polygon P, the first two points define a line while the third point lies left of this line, as does each following point. As two adjacent points define a line segment x, and since two arbitrary following points, which define another line segment y, both lie left of x, so does line segment y; i.e. for each pair of lines x and y of P it holds that $x_y = l$.

Conversely, if $x_y = l$ for each pair of lines of P the same holds for the endpoints of x and y, which defines the convexity on the triangle orientation for three points of P.

□

Definition 4.13 (Convex Hull)
P is a polygon. The convex hull of P, in short $\mathcal{CH}(P)$, is the most minimal simple convex polygon that completely covers P.

Note that line segments of P may lie on the boundary of $\mathcal{CH}(P)$ since P is covered by $\mathcal{CH}(P)$ and not included as a proper part; this is denoted by $x \in \mathcal{CH}(P)$. For such line segments as well as for some other line segments we define:

Definition 4.14 (Global Extremum)
x is a line segment of a polygon P. x is said to be a global extremum of P, for short $\zeta(x)$, if $C(x) = d$ with $d \in \{l, r, F, B\}$.

It then holds that any line segment x of any polygon P, which simultaneously lies on the convex hull of P, is an extremum:

Dependency 4.4
$\forall_{x \in P} : x \in \mathcal{CH}(P) \Rightarrow \zeta(x)$

Proof: $\mathcal{CH}(P)$ is the minimal convex polygon which completely covers P. For convex polygons Dependency 4.3 holds, and all line segments of P which are not part of $\mathcal{CH}(P)$ lie in the interior of $\mathcal{CH}(P)$, i.e. left of any line segment of the anticlockwise oriented convex hull. Therefore,

$$\forall_{x \in \mathcal{CH}(P)} : \forall_{y \in P} : x \neq y \Leftrightarrow x_y = l$$

Taken together with Definition 4.14 $\zeta(x)$ follows.

□

Line segments of $\mathcal{CH}(P)$ which are not part of P behave like extreme line segments with respect to P.

If the course of x is $C(x) = B$ or $C(x) = F$ this course may change from left to right and from right to left; in such cases it holds that x is a global extremum, but it may additionally hold that $x \notin \mathcal{CH}(P)$. This is the reason why Dependency 4.4 is unidirectional.

Another observation concerns the relationship between an extremum and the extent of its course:

Dependency 4.5
$\eta(C(x)) = 1 \Rightarrow \zeta(x)$

Proof: if $\eta(C(x)) = 1$ the rest of the polygon has to be either somewhere left of or right of x, or in front of x or in the back of x, i.e. it then holds that $C(x) \in \{l, r, F, B\}$; together with Definition 4.14 it then holds $\zeta(x)$.

□

Conversely, $\zeta(x)$ does not necessarily imply that $\eta(C(x)) = 1$, but it does hold that:

Dependency 4.6
$\zeta(x) \Rightarrow \eta(C(x)) \leq 5$

Proof: by Definition 4.14 $\zeta(x)$ implies $C(x) \in \{l, r, F, B\}$. For $C(x) \in \{l, r\}$ the course of x runs either completely left of x or right of it; it then holds that $\eta(x) \leq 5$ because there are at the most 5 relations, i.e. $C(x) \subseteq \{F_l, FO_l, D_l, C_l, BO_l\} \vee C(x) \subseteq \{F_r, FO_r, D_r, C_r, BO_r\}$. For $C(x) \in \{F, B\}$ the course of x is completely in the front of x or behind it; in this case it holds that $\eta(x) \leq 3$ because $C(x) \subseteq \{F_l, F_m, F_r\} \vee C(x) \subseteq \{B_l, B_m, B_r\}$. Therefore, the extent of any extreme line segment cannot be greater than 5.

□

Dependency 4.6 is not bidirectional since $\eta(x) \leq 5$ does not necessarily imply $\zeta(x)$. If $C(x) = FO_l F_l F_m F_r FO_r$, for example, the extent is 5, but x is not an extremum. Closely related to Dependency 4.6 is the extent of the convex hull:

Dependency 4.7
$\eta(\mathcal{CH}(P)) \leq 5$

Proof: each line segment x of $\mathcal{CH}(P)$ is an extremum with respect to P: either because it is part of P (Dependency 4.4) or P lies completely left of x (Definition 4.13 and Dependency 4.3). As Dependency 4.6 shows it then holds that the extent of any line segment of $\mathcal{CH}(P)$ must be less than or equal to 5. Consequently, taking the average extent of all courses of the convex hull (Definition 4.10) we obtain $\eta(\mathcal{CH}(P)) \leq 5$.

□

Reversal
In this section two different kinds of changes are distinguished. Supposing a continuous curve, each change from direction x_y to direction x_z corresponds to a change according to the conceptual neighbourhood graph (Figure 4.27), i.e. x_y and x_z are conceptual neighbours. For three or more such adjacent changes we want to distinguish whether the course circulates around the reference line without changing its direction from left to right or right to left, or whether the course includes reversals. Such reversals exist if at least one relation repeatedly appears with one other relation occurring between repetitions, and without completely running around the reference segment. For example, $C_4(x) = F_r FO_r D_r FO_r F_r$ comprises a reversal since the relation FO_r appears twice with the relation D_r in between, and the course has not run around x in the meantime.

Definition 4.15 (Reversal)
x is a line segment of a polygon and $C(x)$ its course. If $C(x)$ comprises at least two sections which run in different directions regarding x, $C(x)$ contains a reversal, and it holds $\varrho(C(x))$.

It can then generally be stated that it holds that:

Dependency 4.8
$\varrho(C(x)) \Leftrightarrow \exists_{u \neq x, v \neq x, w \neq x}(u < v < w \wedge x_u \neq x_v \wedge x_u = x_w \wedge \eta(C(x,u,w)) < 12)$

Proof: u, v, and w are unequal to x, and $u < v < w$ ensures an ordering so that there is a section, s, of the course $C(x)$ which consists of at least three line segments. As it holds that $x_u \neq x_v$ there is at least one change in direction in s while $x_u = x_w$ ensures that the first direction in s is again satisfied at the end of s. The course of s must have been turned backwards in the meantime since it holds that the extent of s, i.e. $\eta(C(x,u,w))$, is less than the extent of the universal scope.

□

Note that $\varrho(C_k(x))$ does not necessarily indicate that $k < 12$ since the course may run completely round x, the course reversing in the meantime at least two times. Examples of courses without reversals from the viewpoint of line x are depicted in Figs. 4.28 and 4.33. The lower part of Figure 4.33 shows the distinction between reversals and changes in the circulation-direction (clockwise or anticlockwise orientation) of courses. In this example it holds that there is no reversal, though $C(x) = FO_r D_r BO_r B_r$ expands to $C(x) = FO_r FO_r FO_r FO_r FO_r D_r D_r BO_r B_r B_r$ and it holds that $\rho(FO_r FO_r FO_r FO_r FO_r D_r D_r BO_r B_r B_r) = rlrlr$. These changes are the consequence of adjacent FO_r-relations. This also shows that reversals form a special case of circulations which we have defined above in the context of scopes. Further specific kinds of circulations could be defined.

Examples of courses with exactly one reversal with respect to x are depicted in Figure 4.34. As the left hand side of Figure 4.29 shows, changes in direction

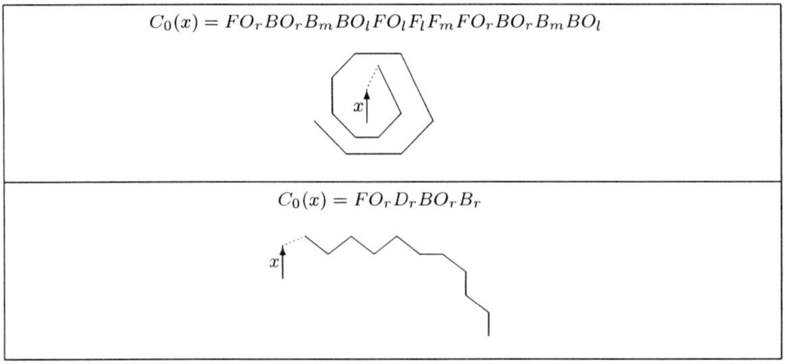

Figure 4.33: Polygons without reversals from the point of view of x

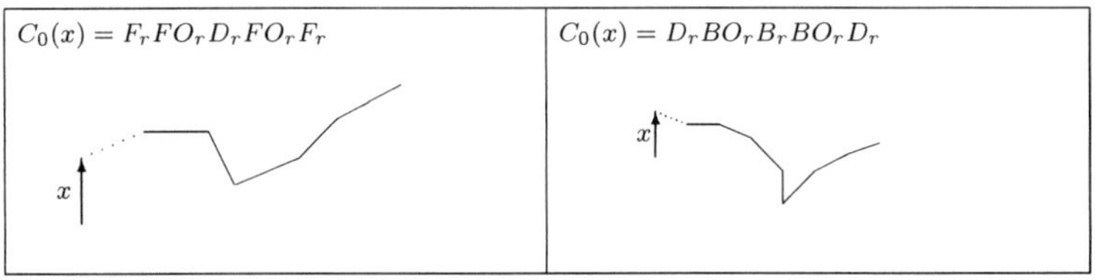

Figure 4.34: Polygons with reversals

of a polygon cannot always be deduced from single courses. It is necessary to test all courses of a polygon, i.e. a polygon contains one reversal only when at least one course of it includes one.

Specific properties

In addition to those global properties which are general in that they are definable without referring to any $\mathcal{BA}_{23}$-relations[5], there are infinitely many conceivable properties which are definable in terms of specific $\mathcal{BA}_{23}$-relations — in fact, each course can be regarded as a specific property.

Bend

The first example concerns two special cases for an extreme line segment, namely

[5]Note that the definition of an extremum refers only to general directions like l, r, F, B, and not to specific $\mathcal{BA}_{23}$-relations.

left-wing oriented and right-wing oriented bends, describing components which frequently occur, having a U-like shape. Assuming a polygon which approximates a smooth curve, a left-wing or right-wing oriented bend is distinguished by a number of courses with the reference segment being an extreme line segment and the rest of the course lying either completely left of the reference segment or right of it.

$$C(x_1) = F_r;\ C(x_k) = B_r Id F_r, k = 2, .., n-1;\ C(x_n) = B_r$$

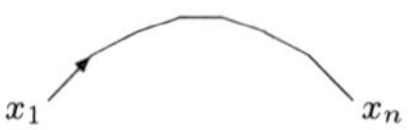

Figure 4.35: A shape with a right-wing oriented bend

Definition 4.16 (Bend)
A polygon which describes a left-wing or right-wing oriented bend is denoted by $\upsilon_l(P)$ *and* $\upsilon_r(P)$*, respectively, and we define*
(a) $\upsilon_l(P) \equiv C(x_1) = F_l \wedge \forall_{k=2}^{n-1} : C(x_k) = B_l Id F_l \wedge C(x_n) = B_l$
(b) $\upsilon_r(P) \equiv C(x_1) = F_r \wedge \forall_{k=2}^{n-1} : C(x_k) = B_r Id F_r \wedge C(x_n) = B_r$

This definition applies also to sections and, as such, a polygon can be characterised by its left-wing and right-wing oriented bends.

The extent of the scope of an intermediate line x_k in a bend is 2 because there are only the relations B_l and F_l, or B_r and F_r involved in a bend, i.e. for $\upsilon_l(P)$ we obtain $\sigma(C(x_k)) = [B_l \varepsilon B_l[\,]F_l \varepsilon F_l]$ and $\eta(C(x_k)) = \eta(B_l) + \eta(F_l) = 1 + 1 = 2$. The curvature of such a polygon is always 1 for an intermediate line and zero otherwise.

Straightness
The second example of a special global property concerns the straightness of polygons. Each course of grade 0, i.e. each arbitrary $C_0(x)$, is locally straight from the point of view of line segment x. Furthermore, there are a number of courses which are also straight, and where the reference line x is not at either end, (i.e. $x_1 < x < x_n$), and these courses are also of grade zero, i.e. $x_1 = x_n^{-1}$. In particular the following courses may be classified to be straight:

$$C_0(x) \in \{(F_l, Id, B_r), (F_m, Id, B_m), (F_r, Id, B_l), (D_l, Id, D_r)\}^6 \qquad (4.20)$$

and the reverse of each. Depending on the length of the section which is straight in a course, one course can be said to be either more or less straight than another.

[6] It is probably useful to define further courses as straight — $F_l Id D_r$, for example.

Definition 4.17 (Straight)
x is a line segment of a polygon and $C(x)$ is its course. $C(x)$ is straight, for short $\iota(C(x))$, if $C(x) = C_0(x)$.

A polygon is globally straight if all its courses are straight. Courses which are not straight are referred to as curved. Curved courses can be compared in order to classify one course as less straight (identically, more curved) than another one, depending on the number of courses which are straight. In accordance with equation 4.5, sections can be defined as straight or curved. Hence, it is possible to characterise polygons by determining their straight and curved sections and by determining which sections are less straight than others. Rather than having two definitions of straightness and curvature, these concepts form a continuous dichotomy, and in addition to determining the straightness of a single polygon it can be determined which of two polygons is less straight.

Note how straightness is related to changes in direction, as defined by Definition 4.7. In the same way that changes in direction only appear if the line segments around the reference line are not inverse to each other, the inverse relation is used in order to define straightness. That is to say, a polygon is straight when there is no change in direction involved.

4.2.3 Summary

Qualitative global properties have been introduced in the current section. These properties are based on the same reference system (section 3.1.1) as the interval relations (chapter 3) and local properties (section 4.1). Local properties concern local orientation information which can be completely described by a fairly small number of relations, namely $\mathcal{BLT}$ and $\mathcal{TLT}$ relations.

By contrast, global orientation information is not so simple to encode. The course of a polygon completely describes its global orientation properties, but the course may be quite large, making it difficult to see its properties directly. This distinguishes local from global properties: there are a fairly small number of local but infinitely many global properties; the former confined to an arrangement of two or three lines, the latter made up of at least four lines but having no upper limit on the number of lines involved. Therefore, a number of characteristics have been defined which describe courses, making explicit global polygonal properties. Relations between these properties have been identified by a number of Dependencies. Table 4.1 summarises the properties which have been discussed.

Having defined different polygonal properties which could apply to particular parts it is useful to relate parts with special properties to other parts. This allows us to distinguish, for instance, two polygons which comprise the same straight or curved sections in the same ordering but are still different from a global perspective. Such parts or sections can be related using $\mathcal{BA}_{23}$-relations. In particular, it is sometimes possible to describe the scope of a course in a concise way, taking coarse relations as introduced in section 4.2.1. Coarse relations can

Table 4.1: Global properties of polygons

Property	Notation	Meaning
Curvature	ξ	number of changes regarding $\mathcal{BA}_{23}$
Scope	σ	range of relations around the reference line
Extent	η	number of $\mathcal{BA}_{13}$-relations which cover σ
Extremum	ζ	$C(x)$ runs wholly left, right, front, or back of x
Bend	υ	P makes up a left- or right-wing oriented bend
Reversal	ϱ	$C(x)$ changes its direction regarding x
Straightness	ι	$C(x)$ is straight

be related by $\mathcal{BA}_{23}$-relations in order to describe their relative position, and as polygonal properties may also be described by local orientation information, i.e. by $\mathcal{BLT}s$ or $\mathcal{TLT}s$, it may also be useful to relate local properties relative to each other with $\mathcal{BA}_{23}$-relations, or even to describe the position of local properties relative to global properties. For instance, having identified for two curves P and Q that there are two extreme segments, $\zeta(x) \wedge \zeta(y)$ in each, we may distinguish P and Q by identifying that $x_y = D_r \wedge y_x = C_l$ for P and that $x_y = F_r \wedge y_x = B_l$ for Q, i.e. P and Q both comprise two extreme segments but these segments are differently related to each other in P and Q.

Let us describe the two polygons in Figure 4.36 by qualitative global properties. We will call the polygon on the left hand side P_1 and the other one P_2. For P_1 and P_2 we obtain the properties described in Table 4.2 and Table 4.3, respectively. This shows that these global properties allow us to distinguish P_1 and P_2, which is not possible by means of local properties: the curvature of P_1 is lower than the curvature of P_2, the scopes of P_1 are open while those of P_2 are all closed, the extent of P_1 is much lower than the extent of P_2 (the latter being more than half the extent of the universal scope), P_1 has two extreme segments but P_2 has none, neither P_1 nor P_2 include bends or reversals, and P_1 is straight by contrast to P_2.

The loop on the right hand side of Figure 4.21 shows an example of a polygon with reversals. $C(z)$ on the left hand side of Figure 4.24 shows an example of a right-wing oriented bend. The set of properties which have been defined may be extended with further more or less specific properties which are already inherent in the $\mathcal{BA}_{23}$-courses. Depending on the given application domain it may be useful to make more of them explicit.

Finally, we must point out that we have restricted ourselves to the consideration of orientation information and positional information in order to characterise polygons qualitatively. The definitions of global properties are further limited to positional information only, leaving out the orientation of line segments as well as their length. In this way, a number of qualitative properties have been defined which are fundamental in that they are confined to one dimension, namely positional information. Further properties, which are more

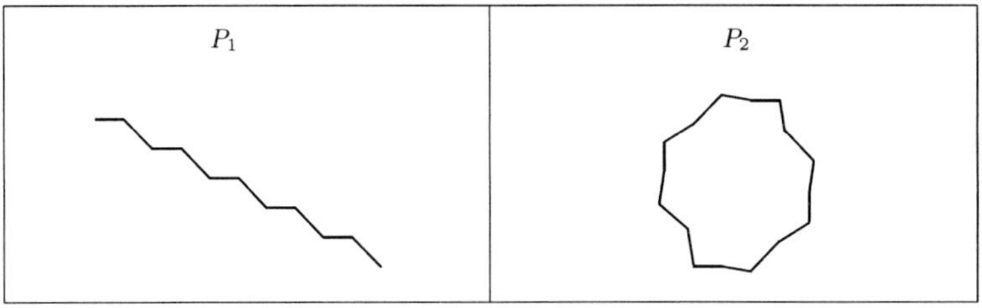

Figure 4.36: Two polygons which are locally equal since they are only made up by $\mathcal{TLT}_6(6)$-relations, but have differing global properties

constrained, can now be defined on the basis of these positional relations, taking into account the orientation and length of line segments.

Table 4.2: Global properties of P_1

P_1
Curvature $\forall_{i=1}^{n} : \xi(C(x_i)) = 0 \Rightarrow \xi(P_1) = \frac{n*0}{n} = 0$
Scope $\sigma(C(x)) = \begin{cases}]F_r, \varepsilon, F_r], & x = x_1 \\ [B_r, \varepsilon, B_r[\ ]F_l, \varepsilon, F_l], & x = x_i,\ i = 2,..,n-1 \wedge even(i) \\ [B_l, \varepsilon, B_l[\ ]F_r, \varepsilon, F_r], & x = x_i,\ i = 2,..,n-1 \wedge odd(i) \\ [B_r, \varepsilon, B_r[, & x = x_n \end{cases}$
Extent $\eta(C(x)) = \begin{cases} 1, & x = x_1 \vee x = x_n \\ 2, & \text{else} \end{cases} \Rightarrow \eta(P_1) = \frac{9}{5} = 1.6$
Extremum $\zeta(C(x_1)) \quad \wedge \quad \forall_{i=2}^{n-1} : \neg\zeta(C(x_i)) \quad \wedge \quad \zeta(C(x_n))$
Bend $\neg\upsilon(P_1)$
Reversal $\neg\varrho(P_1)$
Straightness $\iota(P_1)$

Table 4.3: Global properties of P_2

P_2
Curvature $\forall_{i=1}^{n} : \xi(C(x_i)) = 5 \Rightarrow \xi(P_2) = \frac{5*n}{n} = 5$
Scope $\sigma(C(x)) = \begin{cases}]F_l, r, B_r[, & x = x_i,\ i = 1,..,n \wedge even(i) \\ ]F_r, r, B_l[, & x = x_i,\ i = 1,..,n \wedge odd(i) \end{cases}$
Extent $\forall_{i=1}^{n} : \eta(C(x_i)) = 7 \Rightarrow \eta(P_2) = 7$
Extremum $\forall_{i=1}^{n} : \neg\zeta(C(x_i))$
Bend $\neg\upsilon(P_2)$
Reversal $\neg\varrho(P_2)$
Straightness $\neg\iota(P_2)$

4.3 Combining local and global orientation

Local and global orientation information have been treated separately. Since they describe different aspects, they complement one another and it is time to consider them together. Figure 4.37 shows two polygons which are locally equal and globally different. This can be read off the matrices, the diagonal indicating local properties, i.e. $\mathcal{TLT}s$, the rest of the matrix global properties, i.e. $\mathcal{BA}s$. The polygon on the left hand side is described by the following matrix (in which singularities are ascribed to specific relations):

14	–	F_l	F_l	D_l	B_l	B_l	–
–	14	–	F_l	F_l	D_l	B_l	B_l
B_l	–	14	–	F_l	F_l	D_l	B_l
B_l	B_l	–	14	–	F_l	F_l	D_l
D_l	B_l	B_l	–	14	–	F_l	F_l
F_l	D_l	B_l	B_l	–	14	–	F_l
F_l	F_l	D_l	B_l	B_l	–	14	–
–	F_l	F_l	D_l	B_l	B_l	–	14

For the polygon on the right hand side of Figure 4.37 we obtain:

14	–	F_l	FO_l	D_l	BO_l	B_l	–
–	14	–	F_l	FO_l	D_l	BO_l	B_l
B_l	–	14	–	F_l	FO_l	D_l	BO_l
BO_l	B_l	–	14	–	F_l	FO_l	D_l
D_l	BO_l	B_l	–	14	–	F_l	FO_l
FO_l	D_l	BO_l	B_l	–	14	–	F_l
F_l	FO_l	D_l	BO_l	B_l	–	14	–
–	F_l	FO_l	D_l	BO_l	B_l	–	14

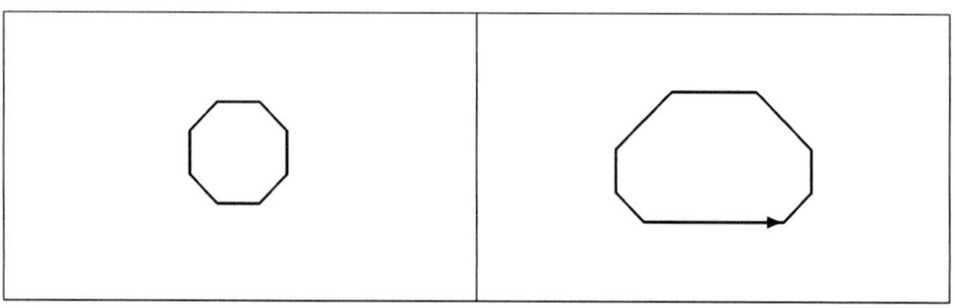

Figure 4.37: Two examples for $\mathcal{TLT}_6(14)$-contours

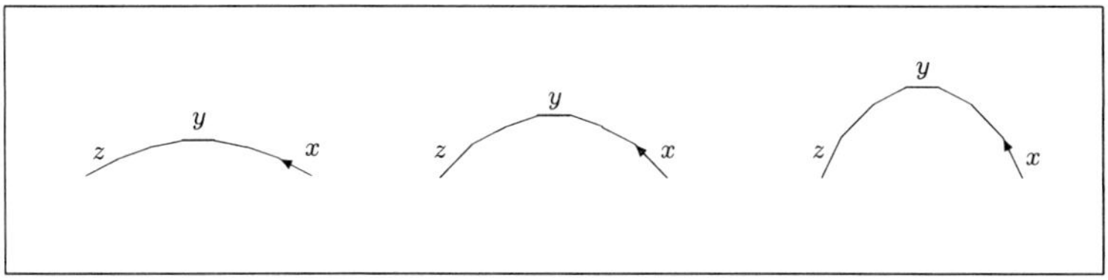

Figure 4.38: Three left-wing oriented bends

When reconciling local and global orientation information it makes sense to use $\mathcal{TLT}_{36}$ because tripartite line tracks fill the gap around the reference line in the $\mathcal{BA}$-matrix, $\mathcal{TLT}_{36}$-relations are oriented like $\mathcal{BA}$s, and the orientation grid is used as a reference system for both $\mathcal{BA}$s and $\mathcal{TLT}$s. In the following we omit the index 36. In order to demonstrate how $\mathcal{TLT}$s and $\mathcal{BA}$s complement each other we will analyse first convex and then concave shapes.

4.3.1 Convex arcs

Convex arcs frequently occur as convex shape parts. Therefore, it is useful to be able to distinguish different kinds of convex arcs qualitatively. Current qualitative shape approaches lack suitable means for distinguishing different convex shapes. (Cohn, 1995) proposes a qualitative shape description which is based on the idea of characterising a shape by its concavities. Each concavity in turn is recursively described by its own concavities. This approach treats all convex shapes as equal. Describing polygons using the triangle orientation of vertices (Schlieder, 1996) is unable to distinguish different convex shapes either. The same holds for the approach of (Galton & Meathrel, 1999) who treat all polygons with the same number of sides as being equal. They propose to annotate their description with indices denoting the relative length of line segments, and this could also be done for angles. This is an obvious possibility, and has also been suggested by others, such as (Jungert, 1993) whose method cannot distinguish more different convexities without this extension than $\mathcal{TLT}s$. But such extensions have the disadvantage that measurements are required in order to obtain quantitative information for the purpose of comparing the size of angles or the length of sides. In some cases (especially sketches) such quantitative measurements are not very robust and it would be preferable, if possible, to discriminate different convex shapes from their qualitative properties alone. This possibility is what we will investigate now. For this purpose, we have to identify which properties convex parts have in common and in which respects such polygons can be changed without loosing their convexity.

Two polygons cannot be distinguished if their line segments are equally

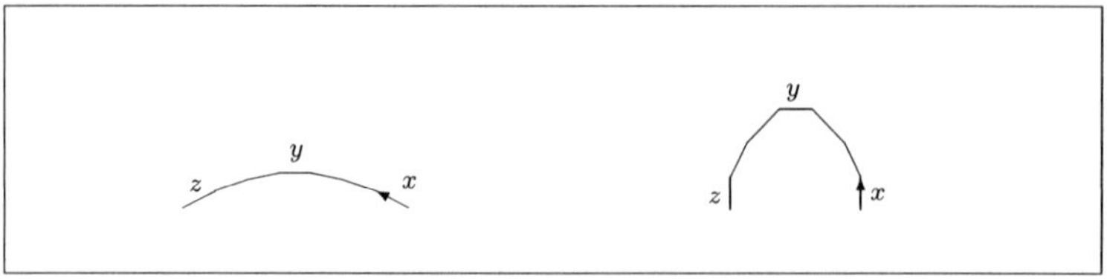

Figure 4.39: Two convex arcs

arranged. It is therefore necessary to consider what kinds of different line arrangements can make up convex polygons. Figure 4.38, for example, shows three convex arcs. They share the same description, that is, they are equal in terms of positional relations, and they all describe left-wing oriented bends. By Definition 4.16, a bend is described by $C(x) = F_l \wedge C(z) = B_l$ and for all intermediate line segments it holds that $C(y) = B_l I d F_l$. As soon as orientation relations between line segments are incorporated the arcs can be distinguished since for the left arc it holds that $x_z^{\phi} = F_l$ whereas for the right one it holds that $x_z^{\phi} = B_l$. The arc in the middle lies exactly at the transition between these two cases.

When we restrict ourselves to positional relations differences between two convex arcs have to be larger in order to distinguish them. For an example see Figure 4.39. For the left polygon it still holds that $x_z = F_l$ but for the right one it now holds that $x_z = D_l$. In order to systematically consider what kinds of convex arcs exist which can be distinguished by positional relations we need only to enumerate all possible convex polygons. Since these examples deal with open polygons, we could consider the start points to be connected to the endpoints in order to be able to apply Definition 4.12. Dependency 4.3 then holds, i.e. every line is left of every other one.

Since we are considering only arcs and not arbitrary convex parts, locally only $\mathcal{TLT}(3)$ relations are used. The simplest convex arcs are shown in Figure 4.40. Their descriptions start with the line segment on the right hand side of each polygon. A third class of arcs is asymmetric, as shown in Figure 4.41. An arbitrary arc can be constructed from more than three line segments. Such arcs either belong to one of those three classes where the order of F_l, FO_l, BO_l, and B_l relations is the same as in the examples of Figs. 4.40 and 4.41, or are more sophisticated polygons constructed from these in the correct order. These follow from the combinations of relations in 4x4-matrices, all of which are shown in Figure 4.42. In addition to the different convex arcs which can be identified by qualitative line arrangements, different closed convex shapes (Figure 4.37) can also be distinguished.

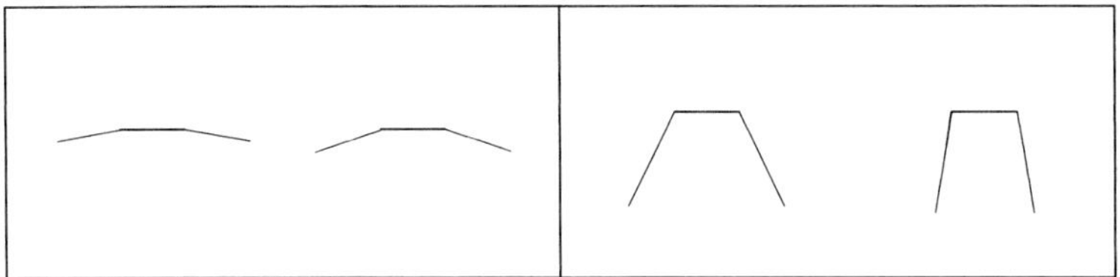

$$\begin{array}{ccc} 3 & - & F_l \\ - & 3 & - \\ B_l & - & 3 \end{array} \qquad \begin{array}{ccc} 3 & - & FO_l \\ - & 3 & - \\ BO_l & - & 3 \end{array}$$

Figure 4.40: Four convex arcs belonging to two different classes

4.3.2 Convex shapes

Having discussed convex arcs we now turn our attention to closed convex shapes. The convex shape with the highest degree of symmetry is a circle. The symmetry of a circle-like polygon can even be deduced from positional relations. The polygon on the left hand side of Figure 4.43 is described by the following matrix where again singularities are ascribed to specific relations:

$$\begin{array}{cccccccc}
14 & - & F_l & F_l & D_l & B_l & B_l & - \\
- & 14 & - & F_l & F_l & D_l & B_l & B_l \\
B_l & - & 14 & - & F_l & F_l & D_l & B_l \\
B_l & B_l & - & 14 & - & F_l & F_l & D_l \\
D_l & B_l & B_l & - & 14 & - & F_l & F_l \\
F_l & D_l & B_l & B_l & - & 14 & - & F_l \\
F_l & F_l & D_l & B_l & B_l & - & 14 & - \\
- & F_l & F_l & D_l & B_l & B_l & - & 14
\end{array}$$

Running anticlockwise around this polygon for each line segment we obtain the following course: $14F_lF_lD_lB_lB_l$. The columns of the matrix show the same pattern indicating the high symmetry of this polygon.

As Dependency 4.3 states, for any two lines x and y of a convex polygon it holds that $x_y = l \in \{F_l, FO_l, C_l, D_l, BO_l, B_l\}$. Therefore, other convex polygons are obtained by using only these relations. To avoid having any concavities

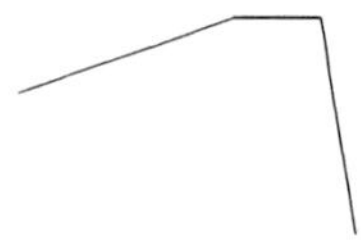

3	–	FO_l
–	3	–
B_l	–	3

Figure 4.41: An asymmetric convex arc

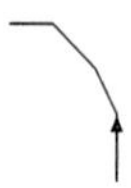

3	–	F_l	F_l
–	3	–	F_l
B_l	–	3	–
B_l	B_l	–	3

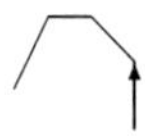

3	–	F_l	FO_l
–	3	–	F_l
B_l	–	3	–
BO_l	B_l	–	3

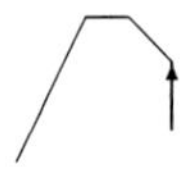

3	–	F_l	C_l
–	3	–	FO_l
B_l	–	3	–
BO_l	B_l	–	3

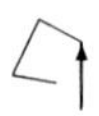

3	–	FO_l	D_l
–	3	–	FO_l
C_l	–	3	–
F_m	FO_l	–	3

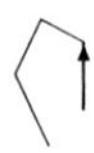

3	–	FO_l	BO_l
–	3	–	C_l
BO_l	–	3	–
D_l	BO_l	–	3

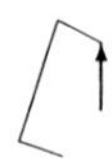

3	–	C_l	B_l
–	3	–	BO_l
D_l	–	3	–
F_l	FO_l	–	3

Figure 4.42: Matrices of convex arcs which are made up of four line segments — all possible variations from the point of view of the first line segment

each course has to take the form $F_l^+ FO_l^* C_l^* D_l^* BO_l^* B_l^+$, meaning that at least one F_l-relation is needed and one B_l-relation, and that a number of overlap-, contains-, and during-relations can be in between the F_l and B_l relations. Varying this pattern of relations we are able to describe different convex shapes. Let us consider how we then distinguish roundish from elliptical shapes.

The elliptical shape on the right hand side of Figure 4.43 is described by:

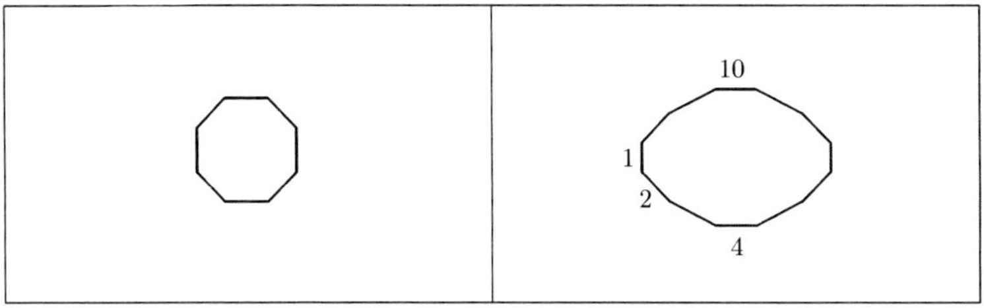

Figure 4.43: A polygonal circle and an ellipsoidal shape

	1	2	3	4	5	6	7	8	9	10	11	12
1	14	–	F_l	F_l	F_l	F_l	D_l	B_l	B_l	B_l	B_l	–
2	–	14	–	F_l	F_l	F_l	F_l	F_l	FO_l	BO_l	B_l	B_l
3	B_l	–	14	–	F_l	F_l	F_l	F_l	FO_l	BO_l	B_l	B_l
4	B_l	B_l	–	14	–	F_l	F_l	F_l	F_l	D_l	B_l	B_l
5	B_l	B_l	B_l	–	14	–	F_l	F_l	F_l	FO_l	BO_l	B_l
6	B_l	B_l	B_l	B_l	–	14	–	F_l	F_l	FO_l	BO_l	B_l
7	D_l	B_l	B_l	B_l	B_l	–	14	–	F_l	F_l	F_l	F_l
8	F_l	F_l	FO_l	BO_l	B_l	B_l	–	14	–	F_l	F_l	F_l
9	F_l	F_l	FO_l	BO_l	B_l	B_l	B_l	–	14	–	F_l	F_l
10	F_l	F_l	F_l	D_l	B_l	B_l	B_l	B_l	–	14	–	F_l
11	F_l	F_l	F_l	FO_l	BO_l	B_l	B_l	B_l	B_l	–	14	–
12	–	F_l	F_l	FO_l	BO_l	B_l	B_l	B_l	B_l	B_l	–	14

By contrast to the symmetrical matrix of the circle the courses of the ellipse differ. For instance, from line segment two to line segment six the number of line segments which are in the front-left of the reference segments decreases while the number of B_l-relations increases. This indicates the asymmetry of the shape from the point of view of different reference lines for which the balance of F_l- and B_l-relations changes.

Another obvious distinction concerns the number of overlap- and during-relations which are above each other in the columns of the matrix. Each column states in which relation one line segment is with respect to all other lines. Line segment four and line segment ten are each in five such overlap- and during-relations, again indicating a kind of asymmetry — that the shape is flatter at these two segments. This can be explained by the fact that a line is more often in an *overlap* or *during* relation when there are a number of lines opposite to it, surrounding it partly from the opposite, thereby inducing orientation grids relative to which the line in question is in *overlap*- or *during*-relation — i.e. if it is situated at a flattened section in a convex, ellipsoidal shape. For line four these line segments are lines 8 to 12. By contrast, line two is not surrounded in such a way but is lying, in a sense, out of the way.

In summary, it is obviously possible to distinguish between symmetrical and asymmetrical convex shapes. Furthermore, asymmetrical convex shapes with flattened sections can be identified. We will now investigate how $\mathcal{TLT}s$ and $\mathcal{BA}s$ can be used together notably qualified in order to describe concave shapes.

4.3.3 Concave shapes

Concavities can be identified as concave by local orientation information, i.e. by $\mathcal{TLT}s$. That is, the set of $\mathcal{TLT}s$ divides into two equivalence classes, one containing concave and the other one containing convex $\mathcal{TLT}s$. The following $\mathcal{TLT}s$ are convex:

$$\mathcal{TLT}(1), \mathcal{TLT}(2), \mathcal{TLT}(3), \mathcal{TLT}(7), \mathcal{TLT}(8), \mathcal{TLT}(9), \mathcal{TLT}(14), \mathcal{TLT}(15)$$

All remaining $\mathcal{TLT}$-relations are concave. Note that a convex shape will consist entirely of convex $\mathcal{TLT}s$, although not every combination of convex $\mathcal{TLT}s$ produces a convex polygon. For example, each closed polygon which contains the convex $\mathcal{TLT}(7)$-relation has at least one concavity.

From this it follows that all concavities of a polygon can be read off the diagonal of its matrix. This allows us, when characterising a polygon, to efficiently find those matrix-entries which are crucial in distinguishing both the concavities it contains and their relative positions. This principal is demonstrated particularly well by the set of shapes depicted in Figure 4.44.

These shapes are all made up of two convex parts and two concavities which are of similar size. (Galton & Meathrel, 1999) use these shapes to demonstrate the limitations of their approach. All these shapes have the same description, namely $\supset\prec\supset\prec$. For the purpose of applying our description to these shapes we have to approximate them with polygons. Using the simple approximation algorithm of (Douglas & Peucker, 1973) we obtain the shapes in Figure 4.45. The differences between the original outlines and their polygonal approximations are shown in Figure 4.46.

The shapes are different in that the arrangements of notches differ. Describing these different arrangements would allow to distinguish these shapes. For the purpose of showing these differences the shapes have been magnified for

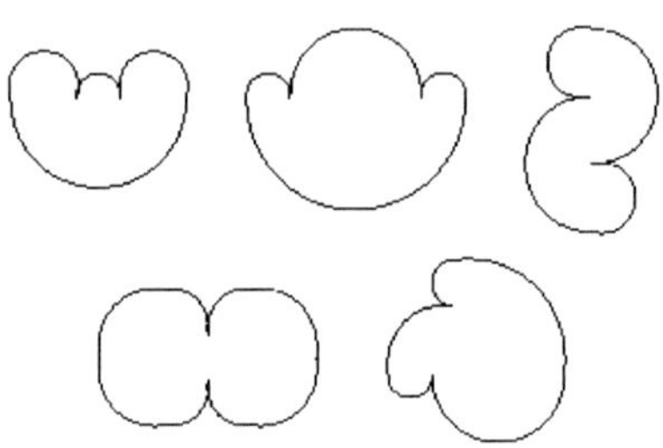

Figure 4.44: Similar concave shapes; Fig. 4 from (Galton and Meathrel, 1999)

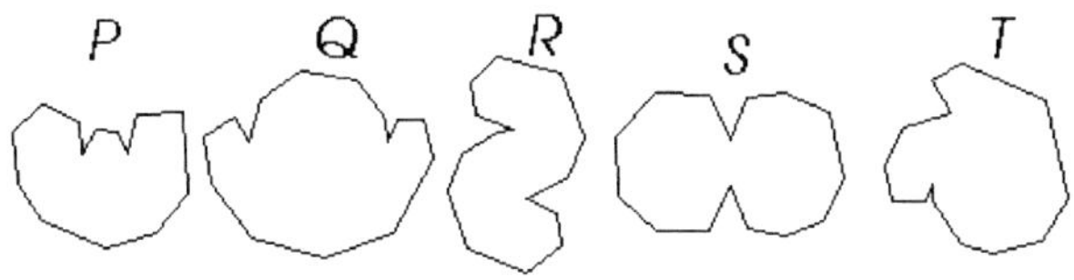

Figure 4.45: The polygonal approximations of the shapes of Figure 4.44

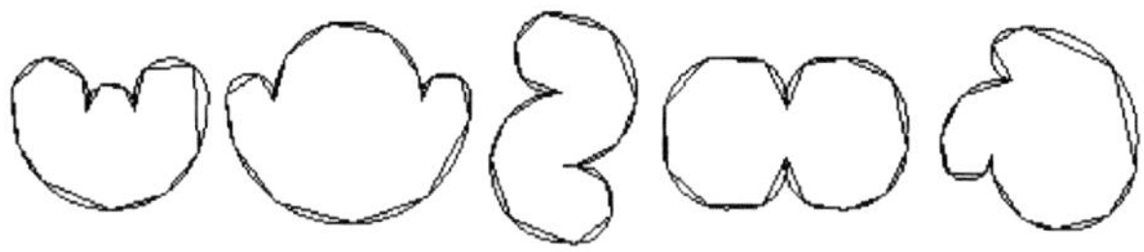

Figure 4.46: The difference between the original outlines and their polygonal approximations, i.e. an overlay of Figure 4.44 and Figure 4.45

Figure 4.47 in order to better display the relative positions of notches.

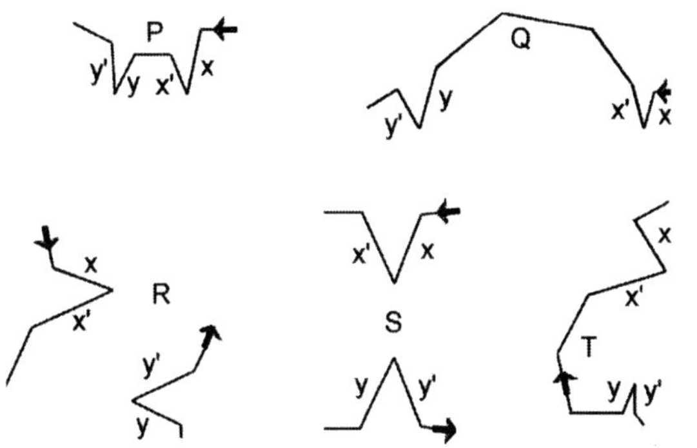

Figure 4.47: The shapes and the arrangements of their notches

In all shapes both concavities are described by $\mathcal{TLT}(5, 27)$ with the exception of Q and T which instead have one concavity with $\mathcal{TLT}(5, 26)$ and $\mathcal{TLT}(11, 27)$, respectively — but note that $\mathcal{TLT}(26)$ is conceptual neighbour of $\mathcal{TLT}(27)$ and $\mathcal{TLT}(11)$ of $\mathcal{TLT}(5)$, meaning that these primitives have significant similarities (see Figure 4.10 and (Gottfried, 2003b)). All convex parts are described by chains of $\mathcal{TLT}(3)$-relations, with the exception of Q and T which also have a $\mathcal{TLT}(2)$ and $\mathcal{TLT}(9)$-relation, respectively, which are both conceptual neighbours of $\mathcal{TLT}(3)$ (see Figure 4.10) and which are also both convex.

For each shape the relative position between the two concavities can be described by taking the lines with the local context of $\mathcal{TLT}(5)$ in place of the concavity (and in one of the two cases of polygon T by $\mathcal{TLT}(11)$). We denote one of the concavities of each shape by x and the other by y. Their relative position is as follows:

$$\begin{array}{llcl} P: & x_y = FO_r & \wedge & y_x = BO_l \\ Q: & x_y = F_r & \wedge & y_x = BO_l \\ R: & x_y = F_r & \wedge & y_x = F_r \\ S: & x_y = F_l & \wedge & y_x = F_l \\ T: & x_y = F_r & \wedge & y_x = F_l \end{array}$$

By this means we may put P, Q, and T in one equivalence class, and R and S in another one. In the first class all concavities are approximately near each other, being on the same side, as far as one is disposed to consider side-like sections in a roundish shape. This derives from their positions: one concavity is left of the other one, whereas the other one is right of the former one. By contrast, for R and S the concavities are opposite to each other, i.e. they are on different sides. This can be derived from their relative positions which are for both points of view equal regarding the left-right dichotomy. In R each concavity is relatively right of the other one, and in S they are left of each other.

R and S can be distinguished as follows. The concavities of R are somehow shifted relative to each other, whereas those of S are opposite to each other, and almost point to each other. Denoting the first $\mathcal{TLT}(27)$-component by x' and the second one by y' (compare Figure 4.47) for R it holds that

$$\begin{pmatrix} x_y & x_{y'} \\ x'_y & x'_{y'} \end{pmatrix} = \begin{pmatrix} F_r & F_r \\ B_l & B_l \end{pmatrix}$$

$$\begin{pmatrix} y_x & y_{x'} \\ y'_x & y'_{x'} \end{pmatrix} = \begin{pmatrix} F_r & F_r \\ BO_l & BO_l \end{pmatrix}$$

The changes of *left* and *right* between the rows reflect the shift of concavities. By contrast for S it holds that

$$\begin{pmatrix} x_y & x_{y'} \\ x'_y & x'_{y'} \end{pmatrix} = \begin{pmatrix} F_l & F_l \\ B_l & B_l \end{pmatrix}$$

$$\begin{pmatrix} y_x & y_{x'} \\ y'_x & y'_{x'} \end{pmatrix} = \begin{pmatrix} F_l & F_l \\ B_l & B_l \end{pmatrix}$$

In this case there is no change between *left* and *right* (compare the orientations of the components of the concavities). Taking these relations for P and Q they cannot be distinguished. But at least T is slightly different from P and Q regarding these relations. This corresponds to the asymmetric curvature of T whereas there is an arc-like convexity in P and Q.

Q and T are quite similar, as both have a small convex part which overhangs relative to the rest of the outline. This can be comprehended very well if rotating

T so that its small convexity is also at the top regarding the picture-plane. The difference in curvature of these convex parts can be derived by the relations which characterise these convexities. Half of them in T are overlap-relations whereas there are only two overlap-relations in the corresponding, less curved, part of Q.

Finally, P and Q can be distinguished by the central convex parts at the top. In P this part is smaller than the one in Q, i.e. it is approximated by fewer lines. Additionally, in P this part is inside the convex hull of the outlines whereas it is lying on the convex hull in Q. This can be derived from its position relative to the rest of the polygon. From the viewpoint of the upper part of the convexity in Q the rest of the outline is completely left of it, indicating that this part lies on the convex hull of the shape. This does not hold for the corresponding part in P which is slightly imbedded in the outline. The same considerations allow us to distinguish P and T.

The described method generalises to arbitrary concave shapes. Recognising concavities by local properties, i.e. by $\mathcal{TLT}s$, these concavities can be related to one another with respect to global properties, i.e. by $\mathcal{BA}s$.

Part III

Evaluation

Chapter 5

Applying qualitative line arrangements

A qualitative representation and calculus for line arrangements has been introduced. In particular, we have investigated how this representation allows qualitative descriptions of polygons, forming a concise means of representing shapes' boundaries. We shall now discuss characteristics of this representation which are related to coarse shape approximations (which maintain only specific global shape properties). We do this by employing our approach to graphically query for images. Such graphical queries are to be sketched by the user who will be looking for objects with specific characteristics. These characteristics are outlined at a coarse granularity level by the user, who is capable of sketching the desired shape only roughly. We will show that qualitative line arrangements allow us to represent these characteristics well enough to use them in querying for similar objects.

Collections of objects d'art, historical tools, or natural objects such as fossilised plants and animals can be found in a wealth of museums and private collections. There are several reasons why people might be looking for such objects: historians investigate the past and track the development of things, art dealers have to serve their customers, designers are looking for stimulation, and so on; there is also the art connoisseur who just appreciates the peculiarity of such objects. In particular, the visual appearance of objects is sometimes important, and may therefore be an appropriate starting point in searching for particular objects. Tools which provide a means for visually querying for objects contained in a collection are then necessary.

From the technical point of view there are two main challenges such a system has to cope with. Firstly, objects which belong to the same category (and which therefore have quite a similar appearance to each other) need to be distinguished. These will, however, obviously differ in some characteristics to which the expert attaches great importance. Secondly, these characteristics are to be specified in a simple but reliable way. There is, necessarily, a trade-off between similarity and variety, the latter relating to those characteristics which are important only to the expert. That is, these characteristics concern those properties that the expert is particularly looking for, and as a consequence which he will take care to specify graphically. It is therefore reasonable to assume that these properties involve perceptually aided distinctions. In this way, we are faced with collections of objects which are similar in that they belong to the same category, but which simultaneously possess perceptually distinct features. In this chapter we will devote our attention to such a collection of objects.

5.1 The Bamberger wax apples

The *Bamberger Naturkundemuseum* owns a two-hundred-year-old collection of malaceous and stone fruits which were brought to Bamberg in 1803 from the secularised monastery Banz by the first curator of the Bamberger Naturkundemuseum, the Benedictine pater Dionysius Linder. Figure 5.1 shows a part of this collection. These wax fruits were manufactured and sold by the *Landes-Industrie-Comptoir* of Bertuch (1747-1824). Friedrich Johann Justin Bertuch was a writer, publisher, and distributer, and owned a number of factories, being a consummate entrepreneur with both commercial and literary ambitions. In the late 18th century Bertuch was both the richest man in Weimar and the largest employer in the region (Kaiser & Seifert, 2000). It is therefore not surprising to encounter the name of Bertuch while studying the products of the 18th century in terms of both objects d'art and commercial products. Bertuch was mainly interested in journal publishing, but his factories produced and sold a variety of things from luxury goods to mass produced art works, including artificial flowers and collections of wax fruits.

The initiator of the production of wax fruits was the priest Johann Volkmar Sickler (1741-1820), a leading pomologist[1]. Sickler edited journals about fruit-growing and orchards, and was author of the journal *Der teutsche Obstgärtner* in which he described the most prominent fruit types. In addition to this, three-dimensional wax models were manufactured (see Figure 5.1) in order to complement drawings and descriptions of different fruits in his and other journals. Similar models have been manufactured by a number of pomologists and companies into the 20th century for the purpose of teaching and advising authorities, tree nurseries, or private circles. The fruits manufactured by the Landes-Industrie-Comptoir which can be found now in the pomological cabinet of the Bamberger Naturkundemuseum are some of the artistically and technically most perfect models (Mäuser, 2002). As they were produced over a period of 19 years, it is assumed that there are many examples still available. Enquiries in Germany and other European countries have been made and some of these models have been found in museums or private collections.

Fruit-growing and pomology reached its summit in the 18th and 19th centuries. It was during this period that a great variety of thousands of types of apples and pears were disseminated. The number of different types available

[1]The field of pomology is a branch of botany which is concerned with fruit-growing.

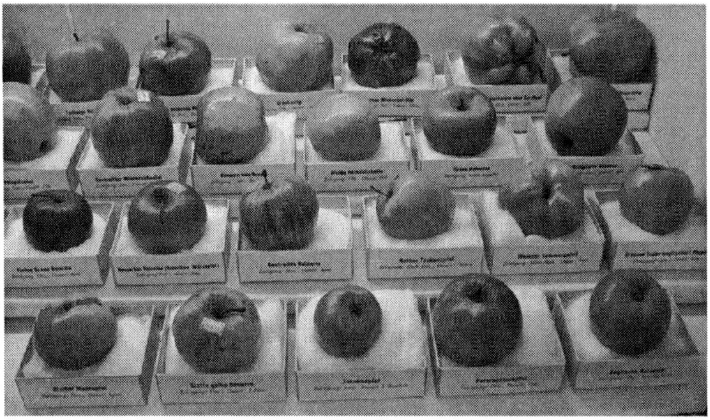

Figure 5.1: The pomological cabinet in the Bamberger Naturkundemuseum

was immense and things were made even more difficult because fruit types have often been given different names in different locations. Extensive papers and books about pomology were written in a number of European countries which tried variously to sort out the confusion, to correctly distinguish varieties of fruit, to assist in fruit-growing, or simply to recommend specific fruits. The lifelike three-dimensional models complemented those books particularly well by showing the actual appearance of fruits.

The fruits of the Bertuch manufacture are hollow and have a thickness of only 2 mm. The material consists of a glaze of beeswax with a dash of *Kremserweiß* (white lead). By contrast to other models which are made up of plaster, papier-mâché, or timber the Bertuch-models are quite sophisticated. The wax has been poured into a two-part hollow body — the plaster cast of a fruit. The wax is then distributed evenly all over the form by shaking the body, which could be detached from the hardened wax fruit later on. The stalks were made from revolved and waxed yarn. Finally, the model was polished and painted. Figure 5.2 shows a damaged model, giving an idea of how the wax apples look from the inside.

Let us put ourselves in place of a pomologist who is interested in apples and their historical development. A collection such as the one in the Bamberger Naturkundemuseum is of great interest to us, since we want to know what kinds of apples existed in the past. The collection is a considerable archive of old fruits; often common in their time, but many of which have subsequently become extinct. For the purpose of identifying old fruits which are rediscovered in nature, the Bertuch-collection is an important, indeed indispensable, source. It is exclusively the visual appearance which is conserved by wax fruits and it is therefore only this to which we have access. That is, we have to specify visual

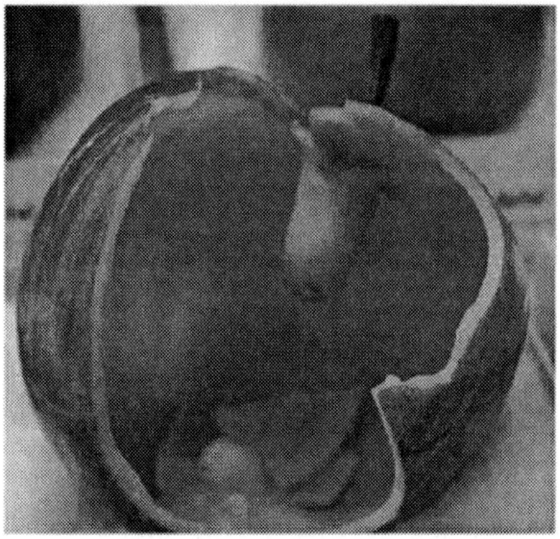

Figure 5.2: A damaged wax apple — *Rother Kronenapfel*

properties of a fruit to identify it. Besides colour and texture, it is primarily the shape of an apple which distinguishes it from others and it is therefore this in which we are interested. For this reason, we will search graphically for apples with specific shape properties. Shapes are rather difficult to specify because the variety of shapes is infinite; therefore, it is necessary to put emphasis on those properties which can readily be made graphically by a human user. That is, we specify shapes using sketches which are to be analysed and evaluated regarding our qualitative description. The description can then be compared with the images in the Bertuch-collection with the object of finding images of apples resembling the graphical query in terms of the necessary shape properties.

5.2 Method

We are interested in the retrieval performance when using graphical queries. That is, we want to know whether our qualitative representation provides an appropriate means for comparing images contained in a collection with the sketch of a shape someone is looking for. One method for evaluating the retrieval performance of search algorithms measures the precision and recall of queries with respect to some reference collection. This method and a number of others are discussed in (Baeza-Yates & Ribeiro-Neto, 1999). In particular, we are then able to evaluate how our approach compares to others.

5.2.1 Performance measure

The Bertuch-collection is our reference collection, which we will denote as C. The set of *all relevant images* regarding a sketched query S is called R_S; we refer to R_S as the *reference set* (of S), and it holds that

$$R_S \subseteq C \tag{5.1}$$

R_S contains all those images which an optimal algorithm would retrieve. The set of images which are actually retrieved by using S is called the *result set* A_S and it holds that

$$A_S \subseteq C \tag{5.2}$$

Those images which are both retrieved by S and relevant regarding S form a set of images R_A which we refer to as the *relevant result set.* It holds that

$$R_A = R_S \cap A_S \tag{5.3}$$

In the worst case R_A is empty, i.e. R_S and A_S are completely different. One obvious case of this is that A_S could be empty, which will happen when no image can be found which matched the query sketch S. In the best case R_A coincides with R_S and it then holds that $A_S \supseteq R_S$. The fraction of the number of all available relevant images, $|R_S|$, contained in the relevant result set, $|R_A|$, is called the *recall* RC of a sketched query S:

$$RC = \frac{|R_A|}{|R_S|} \tag{5.4}$$

A recall is optimal if it is 1, i.e. if $R_A = R_S$. As there can only be as many relevant results as there are relevant images in the whole collection C, it holds that $|R_A| \leq |R_S|$ and as such that $RC \leq 1$. In the worst case it holds that $R_A = \emptyset$ and therefore $|R_A| = 0$ and thus that $RC = 0$. Typically, the recall will be somewhere between 0 and 1, indicating that only a subset of the relevant images have been found.

The recall does not take into account all those images which have been found. It could be the case that $RC = 1$ but that $A_S \supset R_A$ and that therefore $R_S \subset A_S$. In this case, A_S also contains some irrelevant results in addition to all the relevant images. Irrelevant results are considered when measuring the precision of a query, which relates all results obtained to the number of relevant images contained in the result set:

$$PR = \frac{|R_A|}{|A_S|} \tag{5.5}$$

If the result set, A_S, contains all the relevant images *and* no irrelevant images then it holds that $PR = 1$. Sometimes the result set may contain only relevant images but not *all* of them. In this case the precision is still 1, but the recall would be less than 1. The precision decreases when the number of irrelevant results increases since then $|A_S|$ increases while $|R_A|$ does not change. The more wrong images there are in A_S, the lower the precision, even if many (or all) of the relevant images are also found. It holds that $|R_A| \leq |A_S|$ and $0 \leq PR \leq 1$; if there is no relevant image in the result set, it also holds that $PR = 0$. It is assumed that at least one image is found, i.e. that $|A_S| > 0$.

A precision-recall graph can be plotted for each individual query. This graph is derived by comparing the ranked result set A_S with the set of all available relevant images R_S. Let us consider an example. $R_S = \{I_1, I_4, I_5, I_6\}$ and $A_S = (I_4, I_8, I_5, I_9)$, the brackets of A_S indicating a sorted (ranked) set. In this example, the precision-recall graph is obtained as follows: The first image found is I_4. It is contained in R_S and as such it is considered to be relevant. For the first case we consider only the first element in the result set A_S which is the first relevant object, that is $A_S = (I_4)$. It then holds that $R_A = R_S \cap A_S = \{I_4\}$. The recall is then $RC = \frac{|R_A|}{|R_S|} = \frac{1}{4} = 25\%$, with the precision being $PR = \frac{|R_A|}{|A_S|} = \frac{1}{1} = 1$. The next relevant image in A_S is at the third position since I_8 is not contained in R_S, but I_5 is. As such we now consider $A_S = (I_4, I_8, I_5)$. It then holds that $R_A = \{I_4, I_5\}$. It follows that $RC = \frac{|R_A|}{|R_S|} = \frac{2}{4} = 50\%$ and $PR = \frac{|R_A|}{|A_S|} = \frac{2}{3} = 0.\overline{6}$. We proceed in this way until all relevant images in the result set have been considered.

A prerequisite for computing the precision and recall is that we know which images in C are relevant with respect to any query sketch S. For the purpose of evaluating our approach we determine subsets of C. Each such subset represents one R_S, i.e. it contains only those images which are relevant given a specific query S. Admittedly, such a selection is quite subjective. In particular, if the expert who determines relevant sets fails to notice one or more of the relevant images, the resulting R_S will contain only a subset of all relevant images. On the other hand, a reference set which has been identified by an expert contains only those images which the expert regards as relevant. An expert-made reference set, it follows, can be regarded as an appropriate approximation of the *real* set of relevant images in C; in a sense, it may even be regarded as the only appropriate set of relevant images in C — who can decide better than the expert what is relevant?

5.2.2 A qualitative approach

We shall now outline how we analyse both the apples of the Bertuch-collection and graphical queries, i.e. sketches. Gaps in the sketches are closed by morphological operations (Serra, 1983) before contour extraction. The extracted contour can then be approximated by a closed polygon using, for example, the method of (Mitzias & Mertzios, 1994). A granularity level can be defined on the basis of the maximum difference between the original contour and the approximating polygon. The larger this difference is, the coarser the granularity level. Finally, the qualitative description can be derived from the polygon.

Using the methods we introduced in chapter 4, the qualitative matrix of a polygon is analysed (see Definition 4.5). For the purpose of classifying the body of an apple, we determine its overall shape by considering the convex parts of its body which can make up a round, vertical, or wide shape. This can be determined by looking for those columns in the qualitative matrix in which *overlap-* and *during*-relations mount up, showing deviations from a roundish shape and indicating whether these deviations result in a vertical or wide shape regarding the image plane (see section 4.3.2). Concavities can be determined from local properties, i.e. from $\mathcal{TLT}s$ (see section 4.1.2). By this means, a shape can be divided up into its convex and concave parts, which can then be analysed further (section 4.3). This allows us to identify stalks and other uneven patches in the contour, in particular dents and hillocks which show typical properties of different types of objects. Figure 5.3 shows typical examples which can be characterised by patterns of concave $\mathcal{TLT}$-relations, such as chains of $\mathcal{TLT}(4,33)$- or $\mathcal{TLT}(11,26)$-relations. In order to distinguish two kinds of stalks we measure the extent of those parts which have been identified as stalks. As described in Definition 4.10, the extent tells us something about the complexity of a polygon's course, i.e. if primary intervals are placed equally or differently with respect to a reference segment, and to what extent they surround it. This allows us to distinguish stalks which are almost straight from those which are bent.

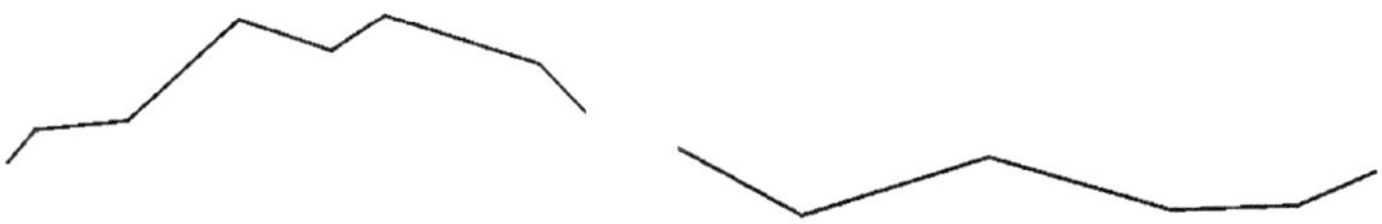

Figure 5.3: Examples for hills at the top (left) and dents at the bottom (right)

5.2.3 A quantitative approach

In order to show how our approach compares to others, it is useful to choose an alternative approach which is also based on polygons. The discussion about the performance of different approaches can then concentrate on the question of how these approaches work on the same representation of an object. Even more important is to compare our approach with an algorithm which is based on the same representation at the same granularity level. Since we have argued that sketches should be analysed at coarse granularity levels in order to focus on necessary features, it should be quite possible to perform better than any approach which is not based on a coarse representation.

By describing and comparing polygons using a classic quantitative geometric approach (specifically, comparing the lengths of, and angles between equivalent line segments in two polygons), we can evaluate our method as an alternative. But before comparing polygons using this quantitative approach, the lengths of the line segments must be normalised with respect to the longest side; this allows to compare polygons with different scales. For two polygons with the same number of vertices, it is then possible to quantitatively calculate their distance from each other. Since there are n ways to match two polygons with n lines, all these matches are calculated, and the one giving the smallest distance is taken as the result. Two equal polygons have a distance of zero, and the more their angles and the lengths of sides differ, the higher their distance becomes.

From now on, we will refer to this approach as the *quantitative approach* and to our own as the *qualitative* one.

5.2.4 Comparing performances

Having two methods, their retrieval performance can be compared using their precision-recall curves for the same query-sketch. In particular, the curves show for which approach the precision decreases faster as the recall increases, and whether the two curves develop similarly or in different fashions. The *precision histogram* shows directly these differences by plotting the difference of the precisions of both algorithms, A and B, for the same recalls:

$$\Delta PR_R = PR_R(A) - PR_R(B) \tag{5.6}$$

with the index R denoting the considered recall. A difference of zero indicates that both algorithms perform equally well; a positive difference shows that algorithm A performs better; a negative difference shows that algorithm B performs better than A.

All these measures for comparing two algorithms can also be used for comparing a single algorithm with a different parametrisation or applied to different reference sets.

5.3 A retrieval experiment

The qualitative approach is now compared experimentally to the quantitative one. For this purpose, a number of sketched queries are made, and a reference set determined for each query. The reference sets can then be taken in order to compare the precision-recall behaviour of the two algorithms.

5.3.1 Image collection

The material consists of a collection of images produced at the Bamberger Naturkundemuseum. These pictures have been taken for the purpose of providing a database which shows fruits from the Bertuch-collection from different perspectives so that an expert can look for specific fruits in the collection.

There are 27 different types of apple in this collection. These apples show various specific properties, such as having a body which is essentially round, vertical-shaped, or wide — a classification which is also used by pomologists (Petzold, 1982). Petzold also describes the surface around the calyx which sometimes shows dents and humps; sometimes there are also hillocks around the stalk, or around both the calyx and the stalk. Besides these properties of the outline, pomologists also consider the shape of the core and its position. But this information is not available from the Bertuch fruits and as a consequence we are restricted to the outlines of the fruits, which Petzold refers to the fruits' relief. It is, however, not important whether we describe exactly the same properties which pomologists use in order to classify fruits. Instead, it is important to identify characteristics someone is able to specify in a graphical query. Such characteristics are then compared with the image collection.

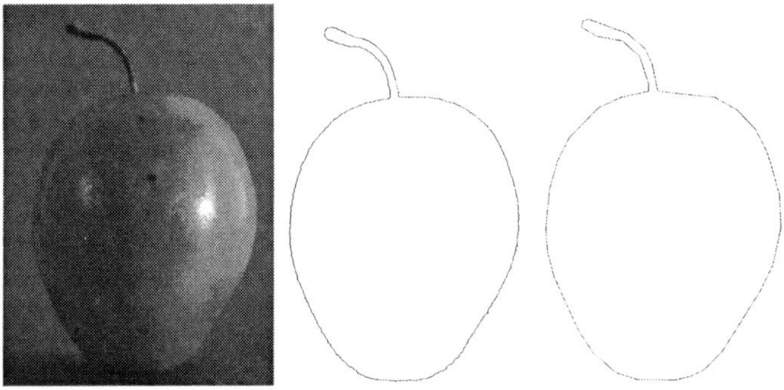

Figure 5.4: An *Italienischer Weisser Rosmarienapfel* (left), its contour (middle), and the contour approximated by a polygon (right)

The images of the collection have been preprocessed by binarising them (reducing them to one bit per pixel). In order to compare the properties of the reliefs, the contours of the fruits are automatically extracted and approximated by polygons. Figure 5.4 shows an example. Finally, the qualitative description for each polygon is generated. Applying the same procedure to the graphical query, a description for this query is generated, and this is compared to the descriptions of the images in the database. That is, our comparison resides completely at an abstract symbolic level at which only necessary shape properties are compared.

Table 5.1 shows the fruits of the collection. Each fruit relief can be specified by the following features:

1. round, vertical-shaped, wide
2. hills at the top or bottom or both or neither
3. dents at the top or bottom or both or neither
4. a straight or bent stalk or no stalk shown

These features combine to a configuration space with $3 * 4 * 4 * 3 = 144$ classes. For example, the *Italienischer Weisser Rosmarienapfel* has a vertical-shaped body and its stalk is bent (see Figure 5.4). Measuring the performance of our approach we shall not distinguish whether dents or hills are at the top or bottom in order to remain invariant with respect to orientation; the user may specify an apple with the stalk at the top or at the bottom — we do not want to place overly restrictive constraints on the user. It is also sufficient to distinguish whether there are dents or not, regardless of their number since the number of dents is not a discriminating factor. As a result of this, the configuration space is reduced to $3 * 2 * 2 * 3 = 36$ classes.

If the search algorithm is unable to find any object in the collection which

Table 5.1: Fruits of the Bertuch-collection and a possible classification; abbreviations: w=wide, r=round, v=vertical-shaped, c=calyx, s=stalk, b=bent stalk; an empty box indicates that there is no such feature

No	Name	Body	Hills	Dents	Stalk
1	Doppelmontagne	v		s	
2	Feigenapfel	w			b
3	Fränkischer Schmeerapfel	v		c	s
4	Fürstenapfel	v			
5	Gedrückter Hartig	w		c	
6	Gestreifter Sommercalvill	w			s
7	Gestreifter Winter Erdbeerapfel	v	c		
8	Grosse Rothe Pilgrim	v		c	s
9	Grosser Pipping	r		c	b
10	Grüne Reinette	w		s	s
11	Grüne Zwetsche	v			b
12	Italienischer Weißer Rosmarienapfel	v			b
13	Neuyorker Reinette	w			b
14	Rother Fenchelapfel	w	s		
15	Rother Herbstcalville	v			s
16	Rother Stettiner	w			s
17	Rother Taubenapfel	v			s
18	Rother Wintercalville	w	s		
19	Schwarzer Borsdorferapfel	w			s
20	Sommer Zuckerapfel	r	c		s
21	Veilchenapfel	r	c		s
22	Weiße Herbstkalville	v	s		
23	Weißer Maatapfel	w			
24	Weisser Sommercalvil	w		c	s
25	Weisser Winterkalvill	v	s		
26	Zwei Jahre dauernde Renette	w			s
27	Zwiebelborsdorferapfel	w			

satisfies all features of the query, it ignores features depending on the priorities which are determined by the domain of apples:

1. The overall shape of the body is most important since it is the most salient feature.
2. The stalk, on the other hand, is less important. If it is not specified in the query then it may be part of objects the user is looking for, or not.
   If a stalk is specified then it matters whether it is straight or bent, assuming that there is one in the result, because the user is obviously looking for a specific stalk.
3. Hills and dents are similar features though hills are more salient than single dents, since they frequently determine the appearance much more than dents (which may be quite small).

From this it follows that the body has the highest priority, followed by hills, dents, and finally the stalk. Taking these priorities into account we are able to generate ranked result sets. Rankings are necessary in order to plot precision-recall graphs.

Some fruits show differences in their relief depending on the viewpoint. For this reason there are two different images for each object, showing it from different viewpoints, making a total of 54 images in the collection. These images, the extracted contours, and their polygonal approximations are listed in Appendix A.

5.3.2 Query-sketches

Six queries from three different people were made; these are shown in Figure 5.5 and (together with their polygonal approximations) in appendix B. To allow just one single granularity level a predefined area was provided, within which one had to sketch the query. As a consequence, the same granularity level works equally well for all queries. In order for this to work, it is assumed that the predefined area is approximately filled by each sketch. Another parameter which is important for determining an appropriate granularity level is the compactness of objects. The less compact an object is, the finer the granularity level of the polygon has to be, so that no details get lost. In our case the objects are all similarly compact as we are faced with one single object category.

5.3.3 Results

Sketch 1

The first sketch is shown on the left hand side of the upper row in Figure 5.5. The reference set consists of seven objects, the positions of which in the ranked result set are shown in Table 5.2, together with the precision and recall. The number following the name of the object denotes one of the two views of that object. Figure 5.6 shows the precision-recall curves for that query-sketch, allowing comparison of the qualitative and quantitative approaches.

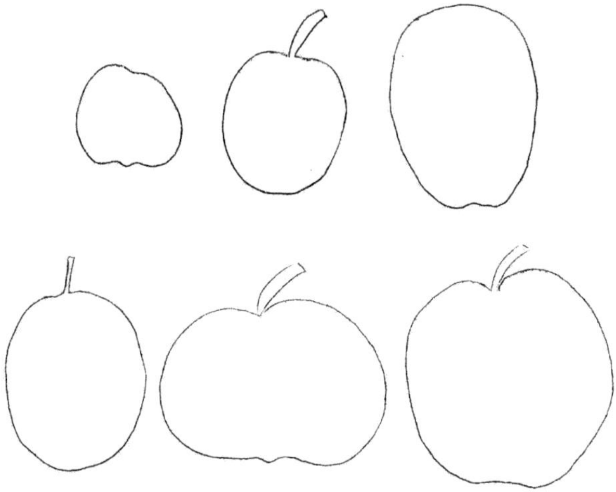

Figure 5.5: Six sketched queries

Our results show that the qualitative algorithm generally performs much better than the quantitative one. While the performance of the quantitative one is almost independent of the recall rate, the qualitative algorithm shows a tendency to achieve higher precision at lower recall rates, with the precision decreasing slightly as the recall increases.

One obvious question is whether the algorithms perform differently at different granularity levels, since we obtain different polygons depending on how finely the fruits' reliefs are approximated. Figure 5.7 shows the performance of both algorithms at a finer granularity level than Figure 5.6. While there is almost no difference for the quantitative approach, the performance of the qualitative one is slightly lower, especially at higher recall rates. Figure 5.8 shows two further precision-recall curves for a coarser granularity level than that used in the first test. While the qualitative algorithm performs significantly better than the quantitative approach at the two finer granularity levels, it is only slightly better at the coarsest granularity level.

The differences in the precisions achieved by the qualitative algorithm depending on the chosen granularity level is shown in the histogram in Figure 5.9. For each recall rate the comparison for the three granularity levels is shown. The algorithm performs better at the medium approximation level than at the fine approximation level except at recall rates between 57% and 71%, and at recall-rates lower than or equal to 28% they work equally well. Both give much better results than the coarsest granularity level unless the recall rates are equal

Table 5.2: The first sketch and its precision and recall (qualitative approach)

Position in ranked result set	Name	Precision	Recall
1	Gestreifter Winter Erdbeerapfel I	1.0	0.14
2	Doppelmontagne I	1.0	0.28
5	Sommer Zuckerapfel I	0.6	0.42
6	Veilchenapfel I	0.66	0.57
7	Veilchenapfel II	0.71	0.71
8	Weisser Winterkalvill I	0.75	0.85
20	Rother Wintercalville I	0.35	1.0

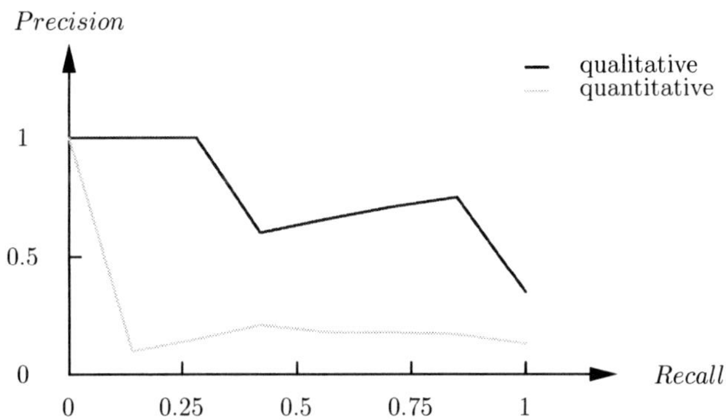

Figure 5.6: The precision-recall curve for the first query

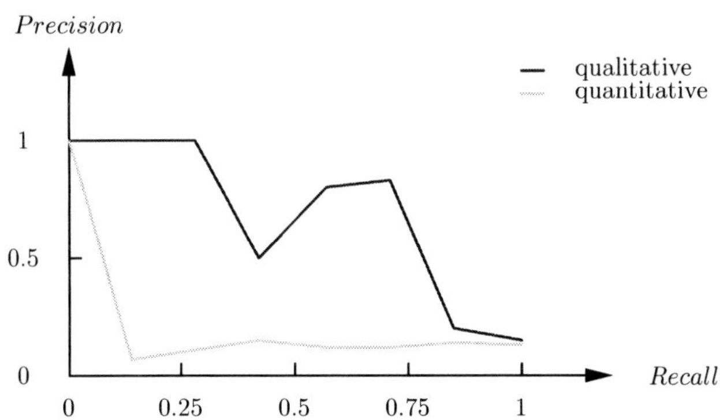

Figure 5.7: The precision-recall curve for the first query at a finer granularity level

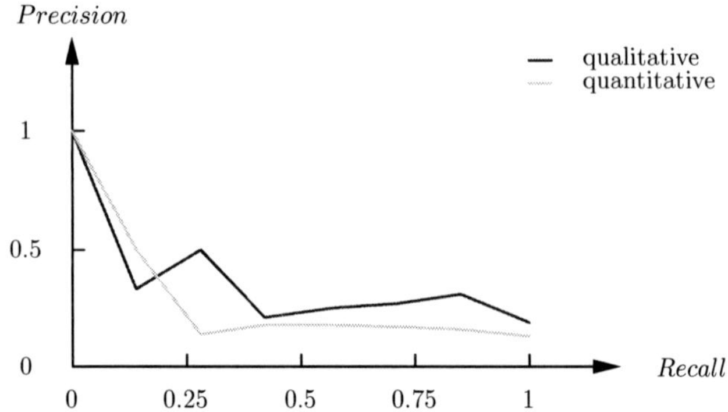

Figure 5.8: The precision-recall curve for the first query at a coarser granularity level

to or higher than 85%, when the algorithm works slightly better at the coarser granularity level.

Query-sketches 1 to 6

Figure 5.10 shows the average precision for all six sketches. Figs. 5.11 to 5.12 show the complete precision-recall curves. Figure 5.11 shows all six query-sketches and their precision-recall. For all sketches the qualitative approach performs better than the quantitative one. However, the result set in the qualitative case has been ranked optimally. That is, if there are a number of equally prioritised results, the best position is always taken. If, instead, we arrange objects in the result set of equally prioritised objects arbitrarily, we obtain the non-optimally ranked precision-recall graph, which is shown in Figure 5.12. Even in this case, however, the qualitative approach performs better. The reference sets of all sketches and the precision-recall rates for both algorithms are given in Appendix B.

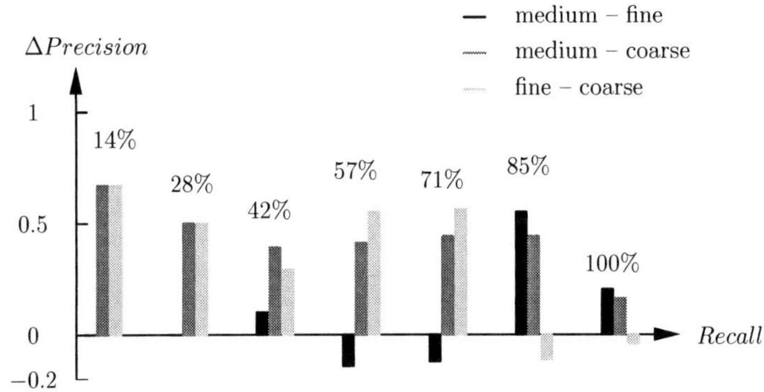

Figure 5.9: The precision histogram for the first query, comparing the three granularity levels with each other

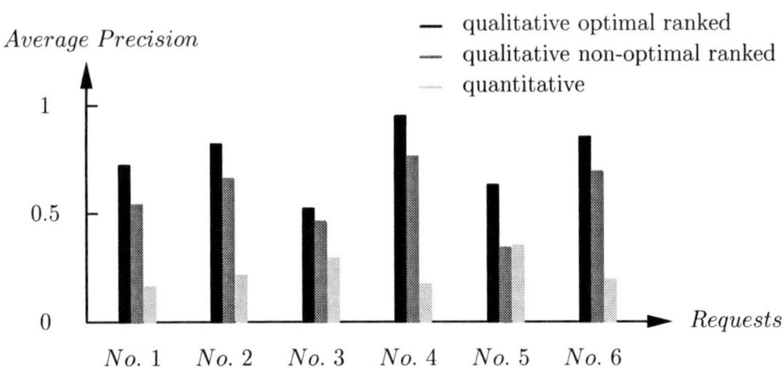

Figure 5.10: The average precision for all six queries and both the qualitative approach and the quantitative one

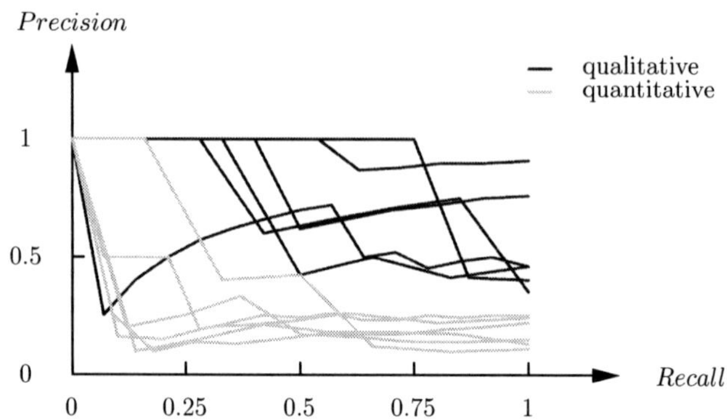

Figure 5.11: The precision-recall curves for all six queries, with the qualitative result ranked optimally

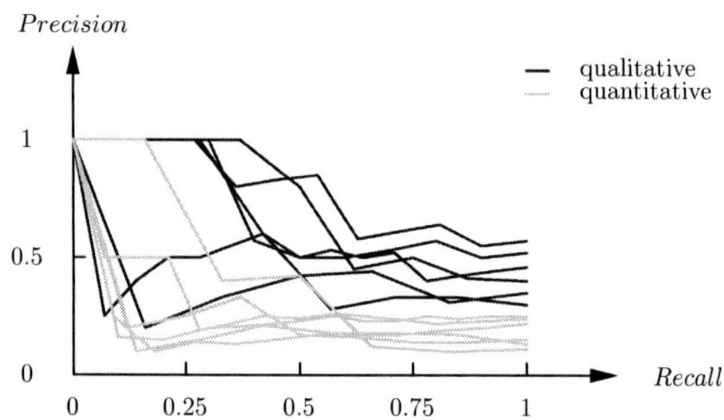

Figure 5.12: The precision-recall curves for all six queries with the qualitative result non-optimally ranked

5.3.4 Discussion

Necessary versus accidental properties

At the beginning, we identified the problem that sketches contain both necessary and accidental shape properties. We proposed to distinguish them by qualitatively describing positional relations between line segments arranged to approximate the underlying shape, assuming that such qualitative relations are closely related to what the human user is able to distinguish both when he is remembering a shape and when he tries to sketch it. By contrast, we assume that accidental shape properties are related to those distinctions which are quantitative. For example, the Grüne Zwetsche and the Italienischer Weißer Rosmarienapfel (left and centre, respectively, in Figure 5.13) are both approximately vertical-shaped and both have a stalk which is relative long and slightly bent. By contrast, an object like the Rother Stettiner (right hand side of Figure 5.13) is wide and has a short stalk. This distinction between short straight stalks and longer ones which are bent is perceptually salient and corresponds to necessary properties. On the other hand, the precise differences between the stalks of the Grüne Zwetsche and the Italienischer Weißer Rosmarienapfel are not important when we only want to distinguish two stalk categories, namely long bent stalks and short straight stalks. The quantitative approach did not make the distinction between accidental properties and necessary properties, but considers details of stalks based on the granularity level used and, as such, regards stalks from the Grüne Zwetsche and the Italienischer Weißer Rosmarienapfel as different. The performance results show that the distinction between necessary and accidental shape properties is relevant to getting appropriate rankings of the results. This supports our thesis that necessary properties are appropriately described by the perceptually aided distinctions which we implemented using qualitative line arrangements while accidental properties are related to metrical distinctions (see our Thesis at the end of section 3.1.1).

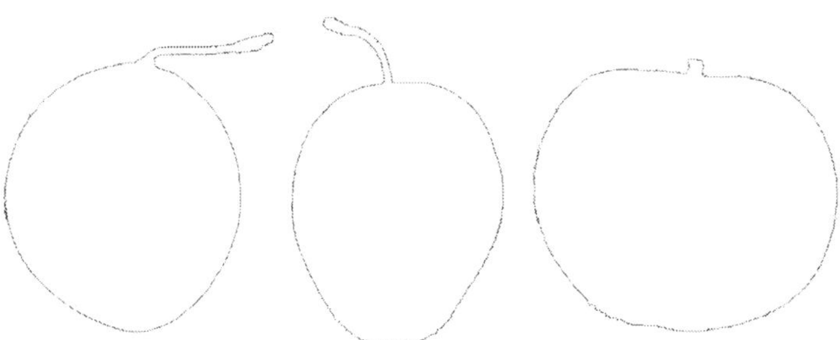

Figure 5.13: The Grüne Zwetsche on the left, the Italienischer Weißer Rosmarienapfel in the middle, and the Rother Stettiner on the right

Qualitative differences

It is obviously much simpler to sketch qualitative differences than precise quantitative distinctions. How does our approach manage to identify the qualitative differences when dealing with the imprecise shapes of the queries? As an example, let us consider the query in Figure 5.14. After closing the gaps (such as the one indicated by the arrow — which is almost invisible at this resolution), the object is binarised, and its contour is extracted and approximated with a polygon (seen on the right hand side of Figure 5.14). As the example shows, there is almost no difference between the original sketch and the polygon, meaning that all necessary properties are still available.

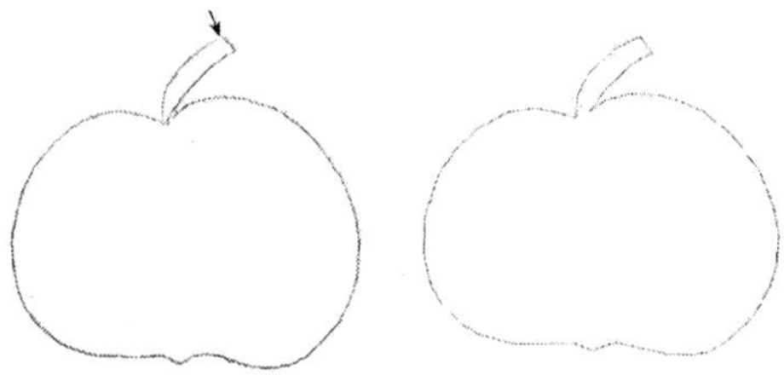

Figure 5.14: A sketched query

We shall have a closer look at the analysis of the stalk in order to demonstrate our method. If we zoom into Figure 5.14 and show the stalk in close-up, we obtain Figure 5.15. Here we see how qualitative line arrangements could allow us to distinguish whether a stalk is bent relative to its body or not. This can be detected by checking whether there are line segments at the tip of the stalk relative to which parts of the body are in relations such as $x_y = BO_r$ and $x_z = FO_r$ — there would not be such overlap relations if the stalk points straight up, rather than being bent in this way. The results of our tests indicate that such relations are robust against the imprecision of sketches.

Yet another variation is shown in Figure 5.16. In this case, the stalks of two apples have been completely detached from their body. On the left is the stalk of the Italienischer Weißer Rosmarienapfel and on the right hand side the one from the fifth sketched query. The extent of each polygon shows whether a stalk is bent or not. Consider the scopes of the polygon on the left hand side:

$$\sigma(C(x)) = [F_r, l, B_l[\ ]F_r, \varepsilon, F_r] \wedge \eta(x) = 7$$
$$\sigma(C(y)) = [F_r, l, B_r[\ ]F_r, \varepsilon, F_r] \wedge \eta(y) = 9$$
$$\sigma(C(z)) = [F_l, l, B_r[\wedge \eta(z) = 7$$

By contrast, for the less bent stalk on the right hand side it holds (for the concave line segments) that:

$$\sigma(C(x')) =]F_r, l, B_l] \wedge \eta(x') = 7$$
$$\sigma(C(y')) =]F_m, \varepsilon, B_m] \wedge \eta(y') = 7$$

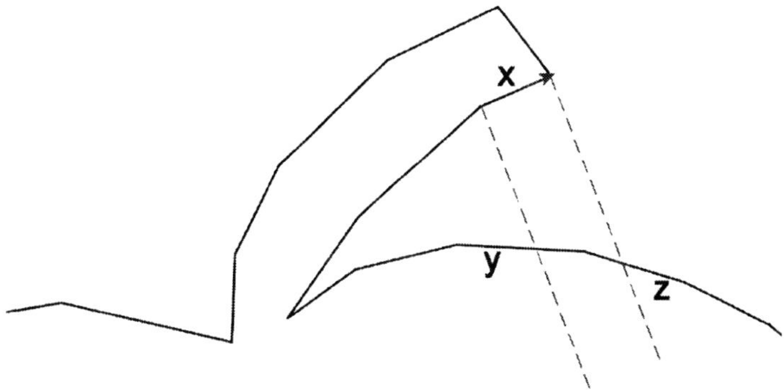

Figure 5.15: The polygonal stalk of a query and its relation to the body

$\sigma(C(z')) = [B_r, \varepsilon, B_r[\,]D_l, l, B_r] \wedge \eta(z') = 5$

The scopes and their extent show some differences. In particular, for the stalk on the left hand side the maximal extent is larger by two than the maximal extent of the stalk on the right hand side. This is actually how we distinguish the two different kinds of stalks in our experiment, and this shows how we focus on qualitative differences of imprecise shapes by describing the arrangements of line segments qualitatively.

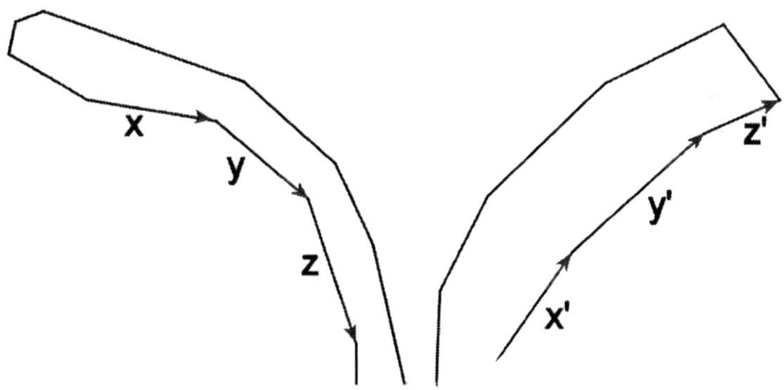

Figure 5.16: Two differently bent stalks and their qualitative line arrangements

Having used a number of qualitative properties which can be derived from the qualitative matrix (see chapter 4), it could be regarded as a drawback of our qualitative approach that one has to determine which of those properties allows us to discriminate properties or categories for the domain at hand. Rather than providing a general similarity measure which is independent of the domain involved (as for the quantitative approach), a multitude of qualitative properties are generically given by the qualitative matrix. In the Bertuch-scenario it was sufficient and not difficult to find a number of appropriate properties to make the distinctions required, but this may well be more difficult in other domains.

Granularity and complexity

Another crucial factor in the performance is the chosen granularity level, as we have seen. Figure 5.17 shows the first query-sketch, with the original contour in the upper-left corner and approximations to this contour at three different granularity levels. The dent at the top disappears at the coarsest granularity level, showing why the algorithm does not perform so well at granularity levels which are so coarse that details disappear. Such details will probably also disappear for some of the objects in the collection and, as a consequence, objects which have different descriptions at finer granularity levels become more similar at coarser granularity levels and eventually get ranked equally in the result set.

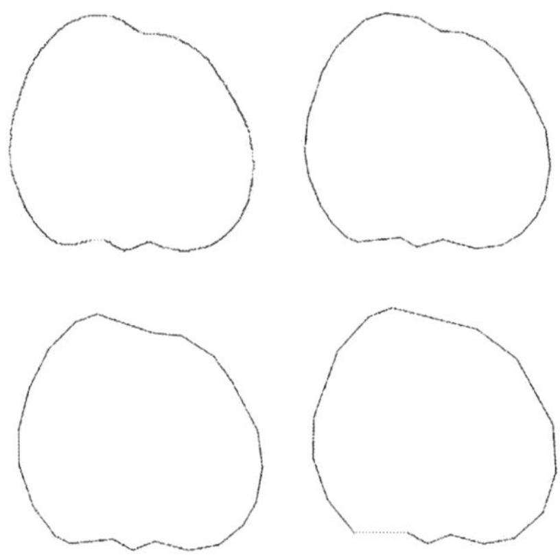

Figure 5.17: The original contour of the first sketch and its approximation at three different granularity levels

We have compared the two search algorithms in terms of their precision-recall behaviour. Another important issue when dealing with image databases is the runtime complexity of search algorithms. The runtime complexity is equal for both approaches. Taking the number of lines involved, n, the quantitative approach needs to try n matches, from which it takes the best (i.e. closest) one. Each match involves linear time complexity, since only the angles of adjacent lines and their lengths need to be computed. As a consequence, the run time complexity of the quantitative approach is $O(n * n) = O(n^2)$. The qualitative approach on the other hand needs only to compare a handful of characteristics, which requires constant time. But the qualitative approach also requires the computation of a polygon's course, which is $O(n^2)$ since the relative position of each line segment is considered in relation to each other line segment, increasing the runtime. However, while the quantitative approach has to be computed every time a comparison is made, the polygon's course only needs to be computed once; the qualitative features of the polygon can be stored and directly referred to whenever its similarity to another polygon needs to be evaluated.

General conclusion

We conclude as follows. In the experiment we investigated whether qualitative line arrangements provide sufficient means for querying an image database using a sketch. The performance results are satisfactory: (i) the first quarter of all relevant images for all but one of the queries were found with a precision of 1.0. The one exception performed at least as well as the other queries at recall-rates from about forty percent upwards. There are sometimes a number of objects which fall into the same category, and as such, are at the same ranking level. If objects at the same ranking level are ranked by chance they are referred to as non-optimally ranked. The performance of non-optimally ranked result sets is slightly inferior to the optimally ranked result sets. (ii) The qualitative approach performs better than the quantitative approach. The sole exception to this is the fifth query, and only when we consider the non-optimally ranked result set. (iii) A single granularity level appears to be sufficient for all objects. That is, we do not need to decide separately for each object how fine it should be approximated in order to achieve satisfactory results.

In addition to these outcomes, attention should be paid to the fact that, in the experiment, we were faced with objects from a single category. This implies that all of the objects are quite similarly shaped and that the method under consideration has to provide means to focus on those properties which are crucial for discrimination purposes. Additionally, these properties may well be sketched quite differently in different queries and especially by different people. As a consequence, a great range of variations has to be taken into account in order to allow for different instances of those properties. Here, we have seen that qualitative line arrangements can cope with these difficulties. The advantages of qualitative line arrangements are twofold: line segments which approximate the underlying shape at a coarse granularity level compensate somewhat for the inexactness of the sketch, and the specific qualitative relations allow quite a large range of variations, compensating even more for the inexactness, while allowing for crucial distinctions. There is a trade-off between those relations which are readily distinguishable in a sketch and those which allow discrimination between different objects or different properties. Here, we have shown that $\mathcal{BA}$-relations form an eligible candidate for such a set of relations.

Chapter 6

Positional-contrast

6.1 Ease of sketching

Graphical queries for the purpose of searching for pictorial information are of growing interest in areas where pictures provide valuable information; for instance, in order to make historical collections of objects d'art available through image databases. Sketching graphical queries is a natural way of revealing the visual appearance of objects one has in mind. The problem which arises is to identify necessary shape properties of sketches, that is, those properties which are not accidental but are necessary for specifying a particular object property. This problem arises in particular with sketches because they are imprecise, and often distorted by the artistic limitations of the sketcher. We face these difficulties by representing necessary shape properties using qualitative line arrangements. In this context, only line arrangements which show properties that are both readily sketched and easily perceivable are considered to be different.

Our approach is based on straight line segments. These encode information about all fundamental dimensions which play a role when representing shapes in pictorial space, namely position, orientation, and length. In the application example with the Bertuch-fruits, we restricted our attention to single shapes, that is to the fruits' outlines. The outline of an object is of particular importance, since it shows how the object's shape is related to other objects, and how it stands out against its environment when being perceived. An outline approximated by a polygon is completely represented by a number of line segments. Positions, orientations, and lengths of the line segments of any polygon are determined by the approximation algorithm, which minimises the difference between the original outline and the approximating polygon. In chapter 4 we introduced a qualitative representation for polygons which allows us to derive a number of characteristics of any polygon. These characteristics are all derived from the way in which line segments are arranged relative to each other. That is, we use neither the relative orientations of line segments, nor their relative length. From the results of the Bertuch-scenario we conclude that the use of

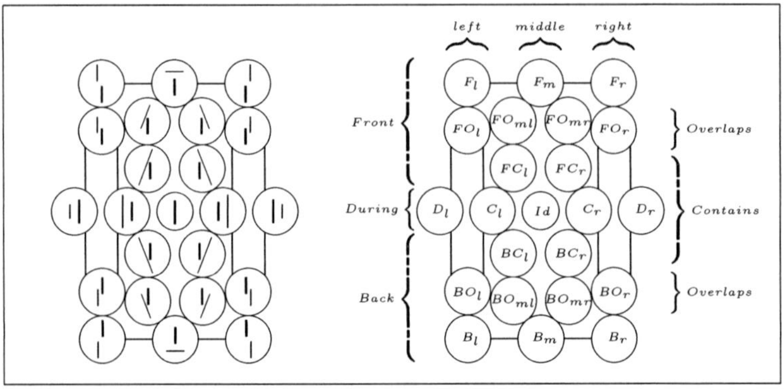

Figure 6.1: Interval relations, defining different positional relations between two line segments

only positional information is, if not sufficient, at least expressive as a means of characterising necessary shape properties in sketches.

From the point of view of sketching, positional relations can be readily distinguished. Consider the basic relations of $\mathcal{BA}_{23}$ (shown in Figure 6.1) — it is certainly not difficult to distinguish, in a sketch, such relations as front-left and front-right, or front-overlap-left and front-left; nor are they difficult to draw. It is primarily the coarseness of the relations of $\mathcal{BA}$ which allows us to cope with imprecise sketches — the correct $\mathcal{BA}$-relation will typically be maintained even in very distorted drawings. In a sketch there are, of course, not only pairs of line segments but arbitrarily curved outlines — but as soon as a curve is translated into a polygon which approximates that curve, curved line segments translate into a number of straight line segments. These line segments represent curved lines, and as such relations between curved parts of the original outline are represented by relations between a number of different line segments. A number of straight line segments coarsely approximate any curved line.

However, in a sketched outline the way in which parts of the outline are arranged relative to each other in order to characterise shapes is crucial. These arrangements describe relative positions of parts. That is, we assume that two outlines in which relative positions of parts are different should differ accordingly, so that their difference can be easily perceived and readily sketched. This difference, or *contrast*, between such two outlines is due to differences in the positional relations of their parts, and we therefore, refer to effects those differences induce as *positional-contrast.* We argue that positional-contrast is both easily perceivable and readily sketchable. As a consequence, it provides a useful means for dealing with imprecise spatial information and characterising necessary shape properties in sketches.

The simplest examples of positional-contrast are the basic bipartite arrangements shown in Figure 6.1. More complex examples are given by the distinctions used in the Bertuch-scenario, which are all derived solely by positional-contrast. Or, consider again the face in Figure 1.2. Stylised drawings such as this face, we argued at the beginning, are made up of a minimum of spatial information, yet this drawing is precise enough for us to determine its category. Since it consists only of a number of lines, the relative positions between these lines and between components of these lines are obviously important in determining that it belongs to the category of faces. Yet another example is Attneave's cat in Figure 2.3. Here, only straight line segments are used, as in the Bertuch-scenario. The cat as a whole is obviously identified by the way in which straight line segments are related to each other, i.e. by positional-contrast.

While we have introduced the basic principles required to establish the concept of positional-contrast further investigations are necessary in order to show how positional-contrast can be applied not only to single objects but to scenes comprised of a number of objects. For example, the course of a polygon P could be defined with respect to another polygon Q, or a third object could be used in order to describe the relative positions between P and Q; given two polygons they could be described relative to each other by using reference parts to show which properties can be perceived from a given point of view. In these cases, the courses obtained do not include the identity relation and, as a consequence, these courses contain no local properties. Especially, concepts such as those discussed in section 3.2.3 are to be considered when investigating positional-contrast in the context of *complex polygons*.

In general, a curve contains properties at different levels of detail. Sometimes it is therefore necessary to consider a number of polygonal approximations of the same curve at different granularity levels. In the Bertuch-scenario a granularity level could be found which worked better than any coarser or finer level, but in domains in which the spectrum of properties (fine and coarse) is larger than in the Bertuch-scenario it may be useful to consider relations between properties at different levels of detail. In this way, the notion of positional-contrast shows its value in simultaneously considering details at different scales.

6.2 Geometrical classification

How does our approach fit into the hierarchy of other geometrical approaches? Qualitative line arrangements impose weaker constraints on geometrical relations than approaches based on affine geometry, since qualitative line arrangements maintain neither angles nor lengths. On the other hand, they allow for stronger constraints than topological approaches such as $RCC8$ which classifies all of the $\mathcal{BA}$-relations as equal, since they are all disconnected.

Instead, the spatial precision of qualitative line arrangements is on the level of ordinal information (projections). A reference interval introduces a system with three oriented lines — the orientation grid. The endpoints of the primary interval are fixed in relation to these lines: each endpoint of the primary line can lie on either of the sides of each line. Two of the lines of the orientation grid are parallel, with the third one aligned orthogonally to them. Therefore, a resolution of 90° angles is considered. For a full discussion of this, see section 2.2.

The distinction from other similar approaches becomes clear when comparing it with the most general approach which is also based on line segment relations in the plane. (Schlieder, 1995b) defines a system where only one line is determined by the reference interval. The positions of the endpoints of the primary interval can then be determined with regard to this line. That is, they are either on one side or the other of that line or else lying precisely on it. In this way, 180° angles are distinguished, leading to a coarser classification of line arrangements than in our case. For a thorough comparison see (Gottfried, 2003b).

6.3 Applications beyond sketching

Before we finish, we will consider further application areas in which the concepts introduced in this work may be of particular interest. Before we leave the issue of sketching we will consider one final example. In the context of image retrieval systems, what matters is that database images match a query image as closely as possible. In contrast, for special domains there are relationships which are particularly important, and the image retrieval system should accordingly focus on such relations rather than on precise correspondences between query image and retrieved images. For instance, in geographical information systems what matters are topological relations between geographical objects (Egenhofer, 1997). It may, for example, be crucial to a particular query that there is a forest and that there is a river which is not connected to the forest; but it does not matter at all what the boundary of the forest looks like, or how far the river is from the forest provided that they are not in contact; such geometrical relationships are not important when we are interested in those images where only the given topological relationships hold. Precise correspondences would retrieve fewer results than there are actually in the image database. But sometimes topological relationships do not sufficiently characterise the query image. For example, it might be crucial to take the curve progression of the river into account. In this case, a $\mathcal{TLT}$ description could be used in order to describe coarsely the curvature of the river in the sense defined in (Gottfried, 2003b). Curvature information is important for many kinds of geographical and artificial objects, including among others coastlines, borders of countries and other regions, rivers, transportation networks (roads and railways), irrigation networks, and sewer systems.

It is desirable that user interfaces become more natural. Personal digital assistants, for example, require natural interfaces for working with them to be efficient. Graphical gestures which are made with pen-like input devices allow for concisely formulated commands. Such gestures need to be both easily remembered and simple to make, but simultaneously there is the need for a large number of such gestures in order to cope with a lot of different situations. The investigations of (Long Jr., Landay & Rowe, 1997) showed that users appreciate gesture-interfaces and demand applications which support more gestures, since they are efficient to use. However, problems arise on the side of the users in memorising gestures, and on the side of the system in correctly recognising them. $\mathcal{BA}s$ and $\mathcal{TLT}s$ may represent a vocabulary of gestures which can be distinguished quite well, and which can be drawn easily.

In the context of cartography we are faced with the problem of dealing with complex maps. Maps often provide the user a vast number of possible ways in which one might get from one place to another one. It is desirable to know the best way in terms of length, comfort, or simplicity. The latter may be determined automatically using $\mathcal{TLT}s$ to determine which of the available paths is the least curved in the sense of qualitative degrees of curvature (Gottfried, 2003b). Such a route should be easily memorised.

Activity patterns represent a further problem class for which characterisation by coarse shape information is useful. Depending on the spatial scale, an appropriate abstraction of precise movement information is necessary. Rather than considering precise movements it may be convenient to choose an abstraction level that allows deduction of typical behaviour patterns, for example, to distinguish explorative versus single-minded activity patterns. This is of interest in several fields, such as aerial navigation routes, satellite orbits, and in investigations into the movements of people and animals. Related to this field are dynamic scenes containing a number of objects, rather than single trajectories. Traffic scenarios and other fields in which a number of autonomous objects interact require some means of representation and reasoning. Problems which arise here are discussed in (Gottfried, 2004a).

A quite different area is the field of trend curves. When considering trend curves we are interested in the overall change of the curve, that is, we are interested in the curve taken by and large rather than in precise properties which are obtainable by curve sketching. Predictions of future trends in share prices and climatic variations are typical examples. (Okabe & Masuyama, 2001), for example, proposed an approach for dealing qualitatively with trend curves. But their approach mainly focuses on consideration of the maxima and minima of a curve; the curve progression is not taken into account. This could be accomplished with different $\mathcal{TLT}$-relations which represent a number of different slopes qualitatively.

Spatial configuration tasks, as applied in interior design or graphical user interfaces, are faced with a set of constraints between objects which have to be satisfied. In contrast to applications such as yellow page layout where primarily the arrangement of equally oriented rectangles is required (Schlieder & Hagen, 2000), objects that can be oriented arbitrarily, such as furniture, require a more complex representation. $\mathcal{BA}s$ provide a means of representing arbitrarily oriented objects.

For all these applications shape information is necessary, but rather than precise geometrical shape descriptions only concise shape characterisations matter in order to solve particular problems. In the context of these applications, it is crucial to determine which shape properties are necessary and which are accidental. Qualitative line arrangements provide a means of characterising necessary shape properties. On the one hand, we propose several characteristics for describing polygons (and thereby shapes) qualitatively. On the other hand, we provide a relation algebra which allows us to deal with arbitrary patterns of line arrangements within the plane.

Part IV

Appendices

Appendix A

Bertuch-collection

Each of the 54 images of the Bertuch-collection, which are used in the evaluation scenario described in chapter 5, is shown, together with its extracted contour and polygonal approximation. The maximal error between the original contour and the polygonal shape is 5 pixels. Note that each apple is shown from two different viewpoints.

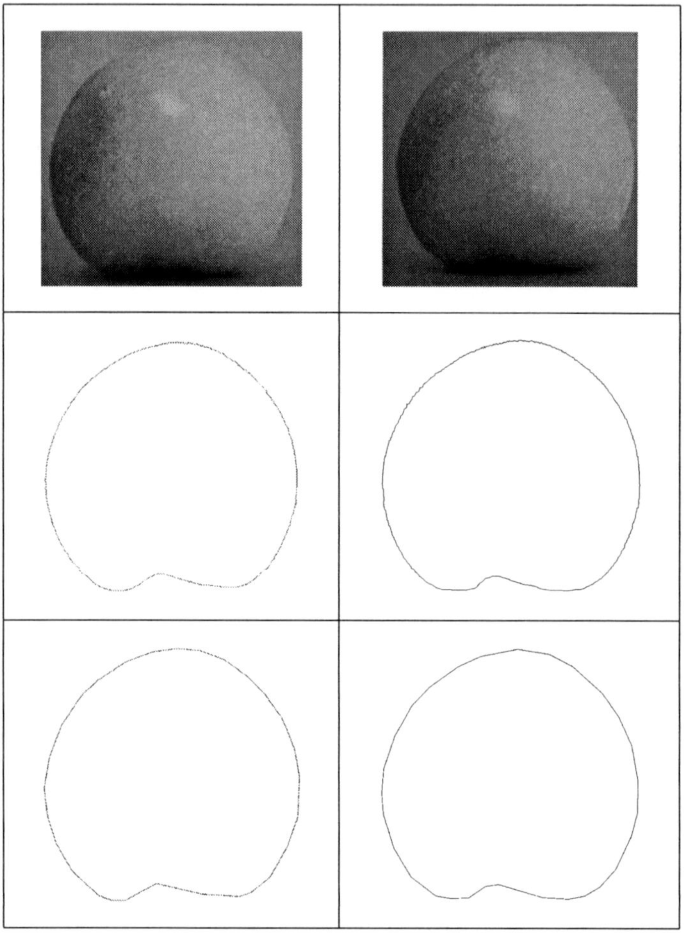

Table A.1: Doppelmontagne

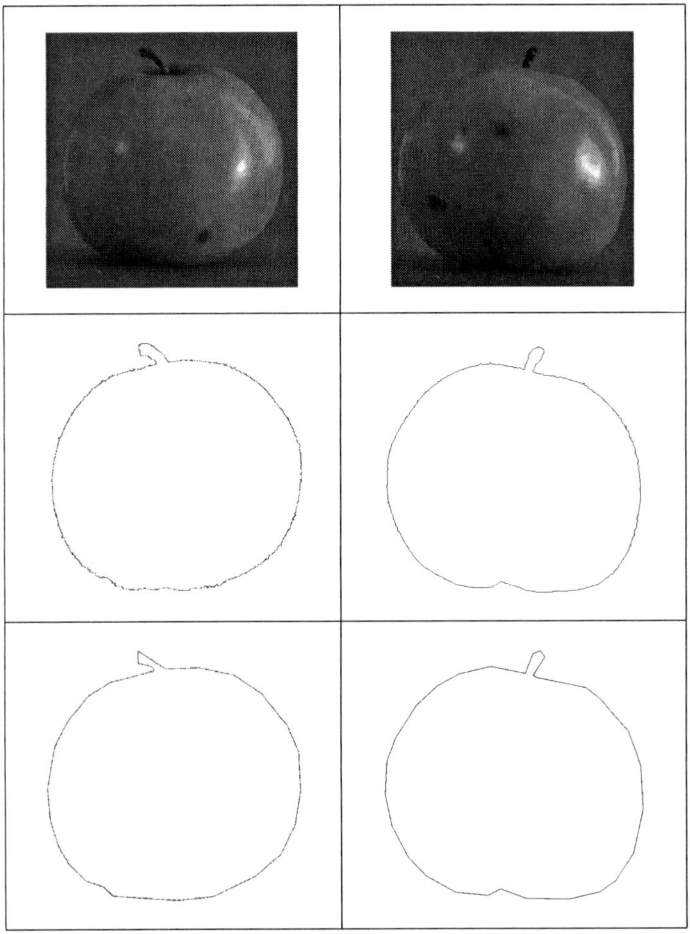

Table A.2: Feigenapfel

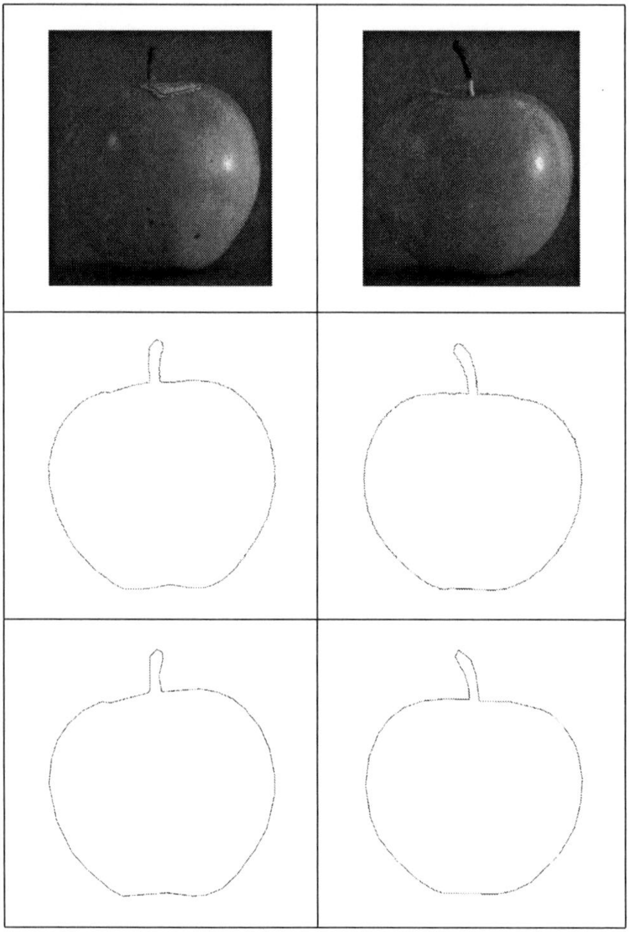

Table A.3: Fränkischer Schmeerapfel

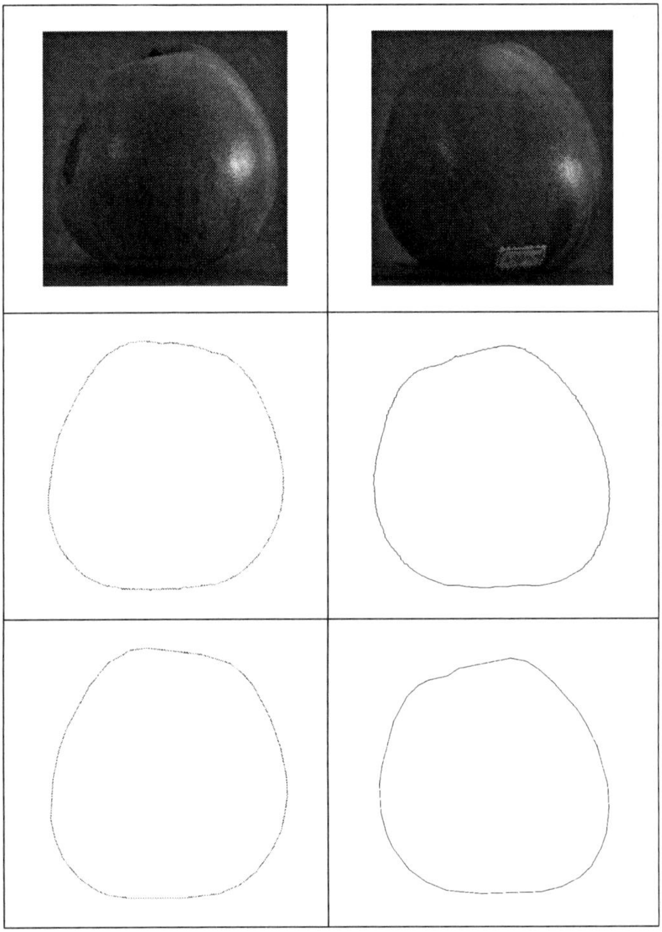

Table A.4: Fürstenapfel

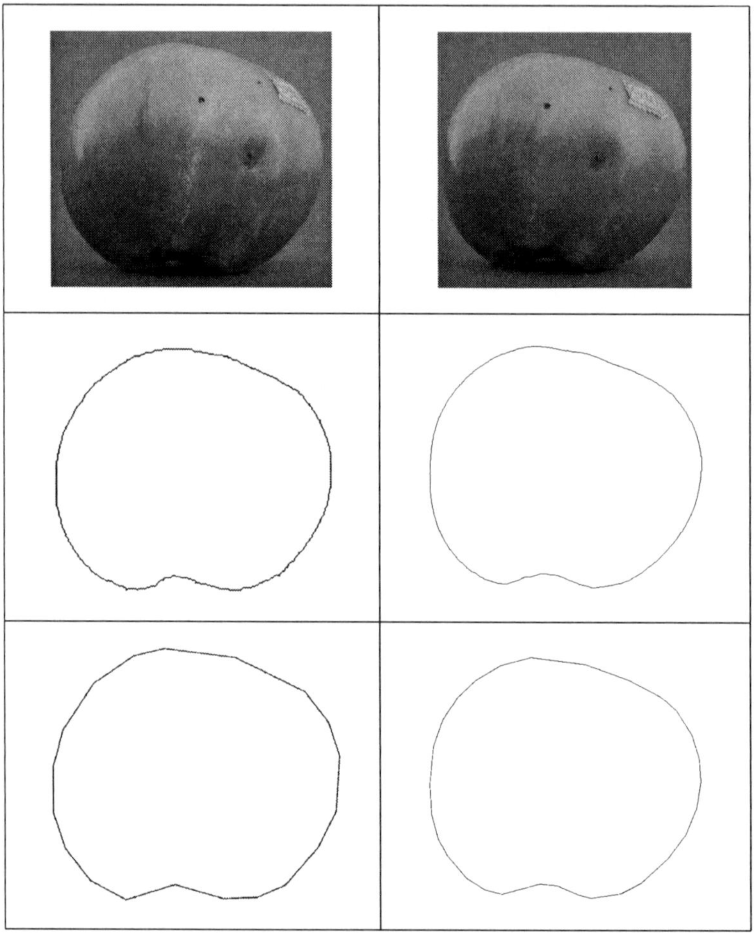

Table A.5: Gedrückter Hartig

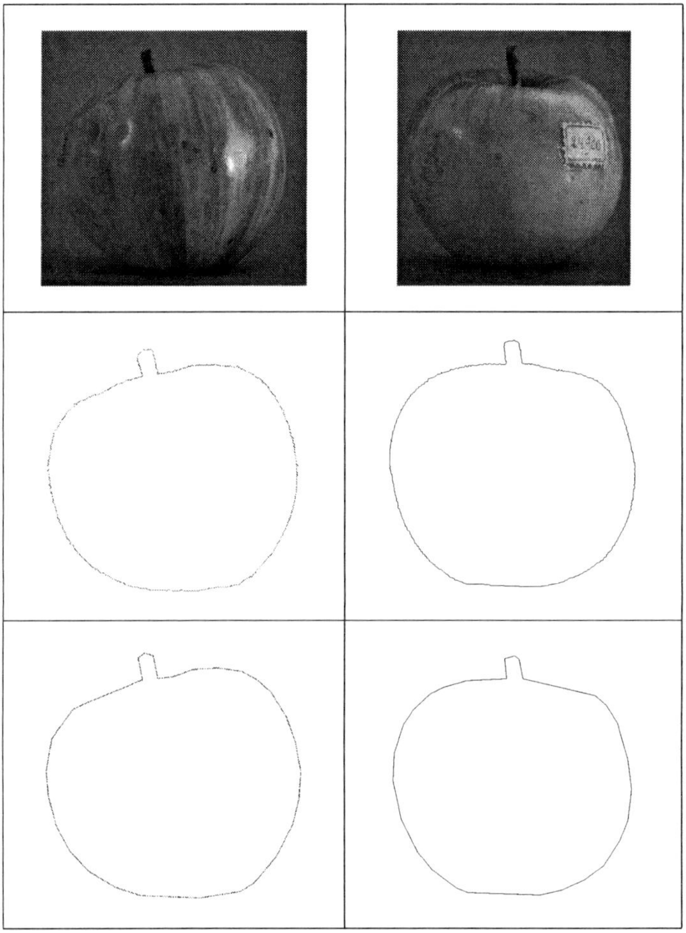

Table A.6: Gestreifter Sommercalvill

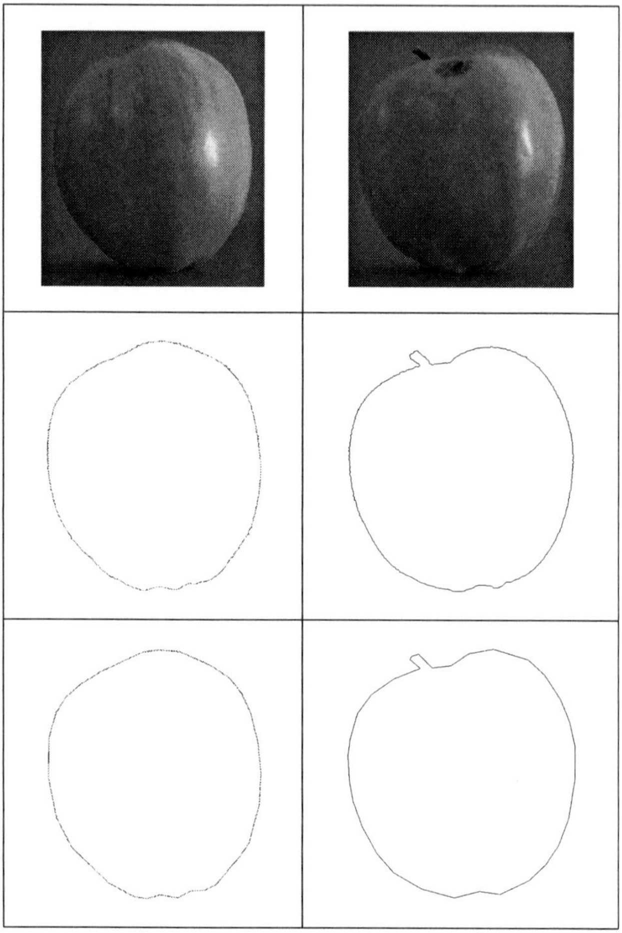

Table A.7: Gestreifter Winter Erdbeerapfel

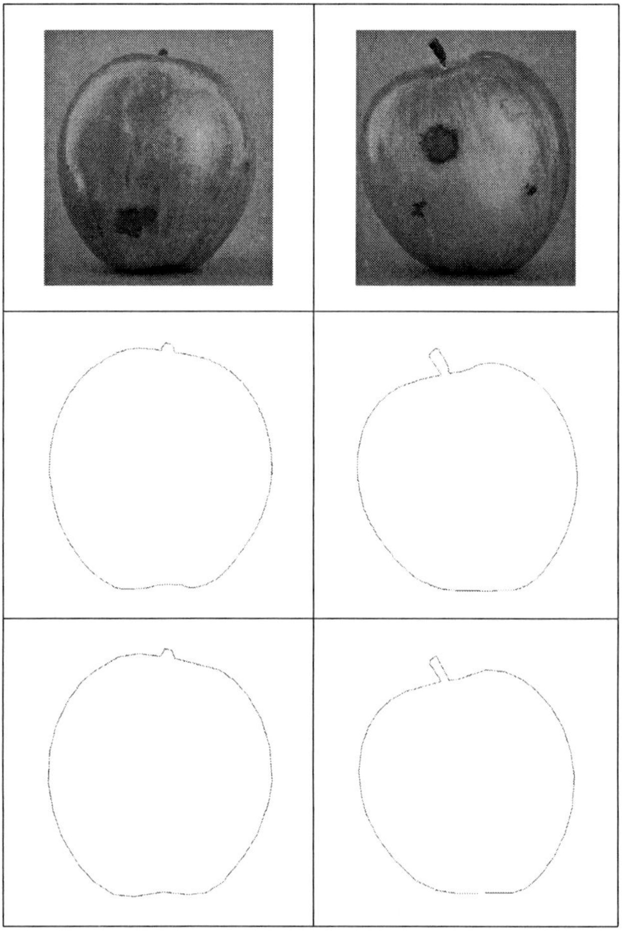

Table A.8: Grosse Rothe Pilgrim

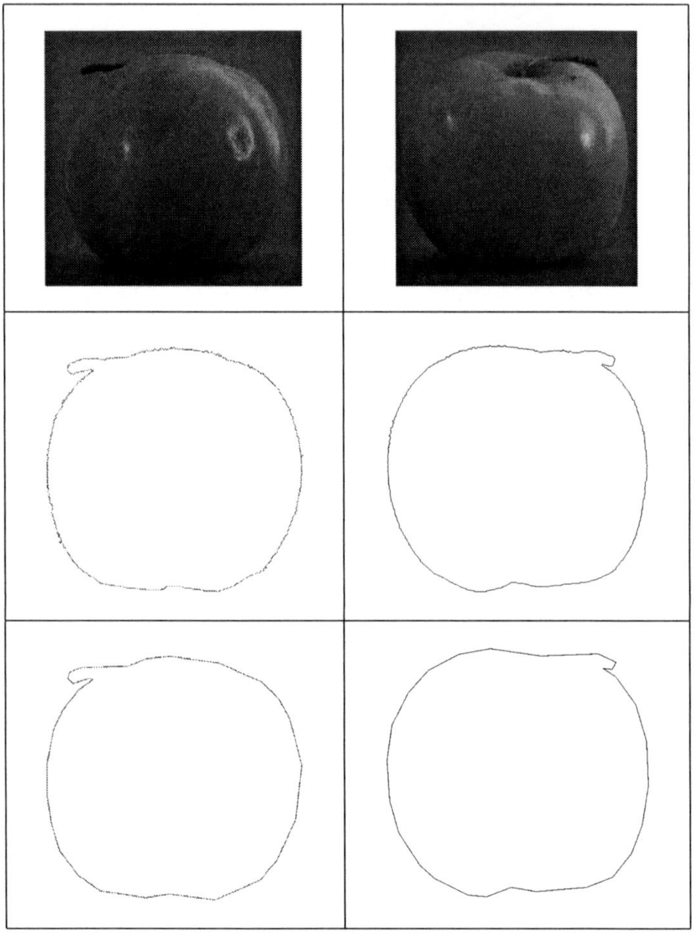

Table A.9: Grosser Pipping

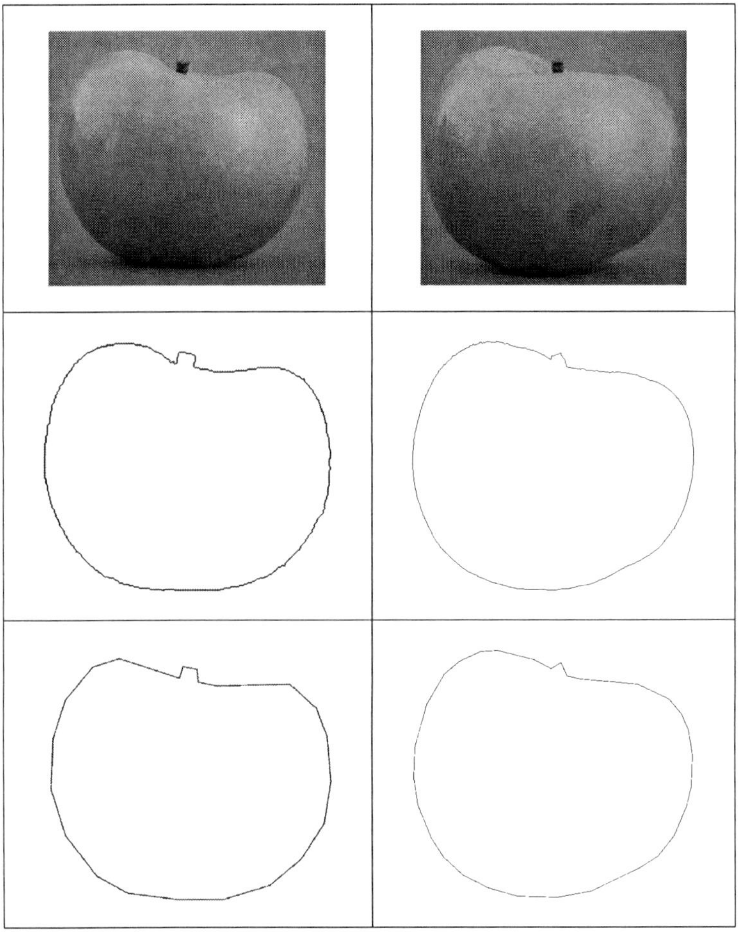

Table A.10: Grüne Reinette

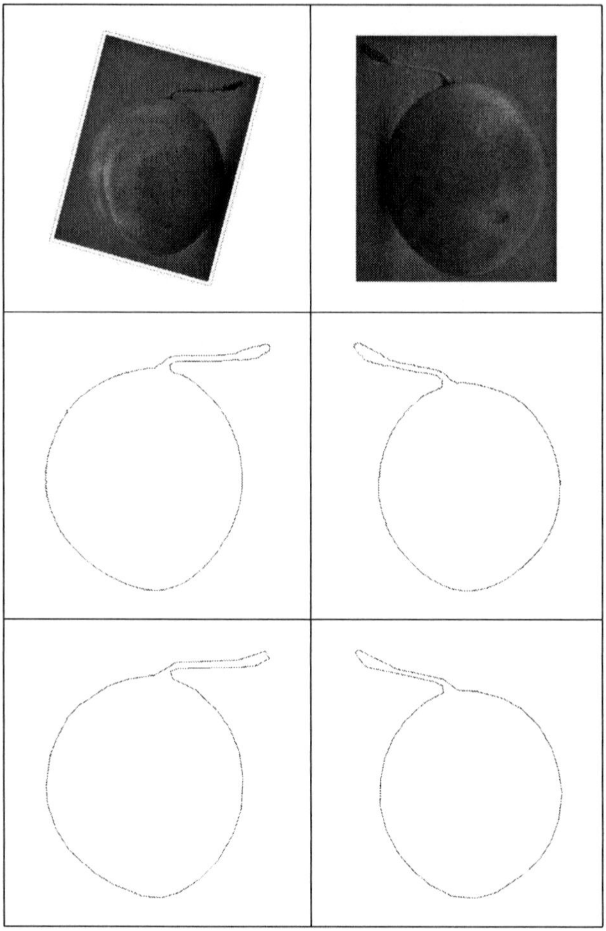

Table A.11: Grüne Zwetsche

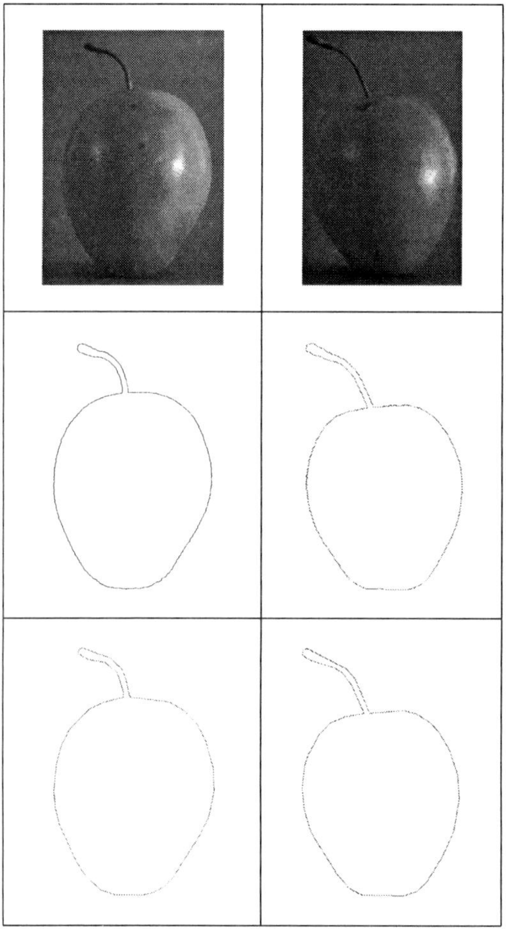

Table A.12: Italienischer Weißer Rosmarienapfel

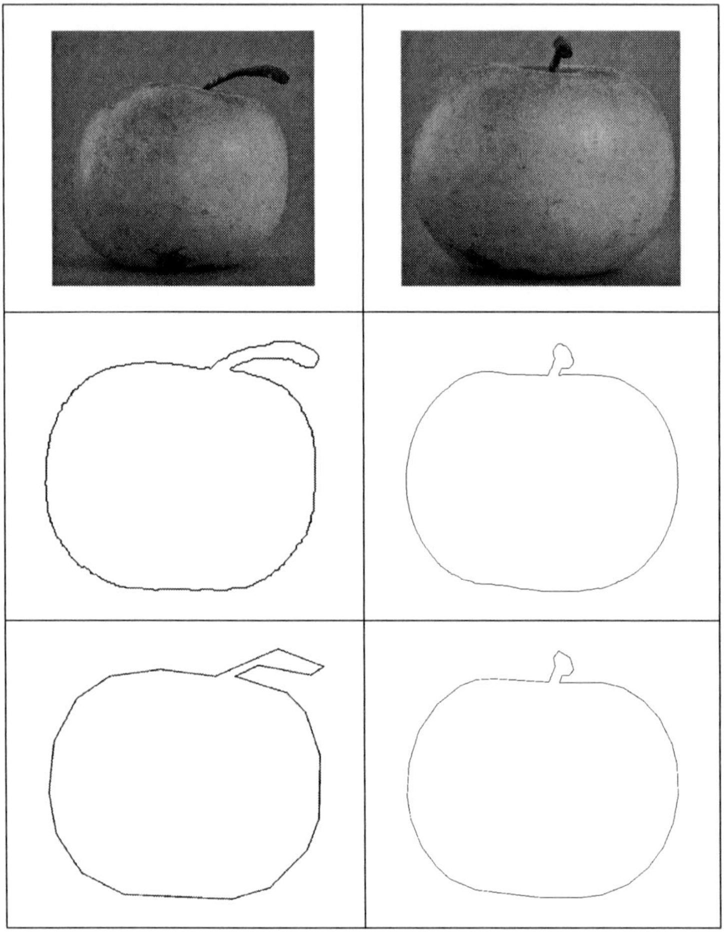

Table A.13: Neuyorker Reinette

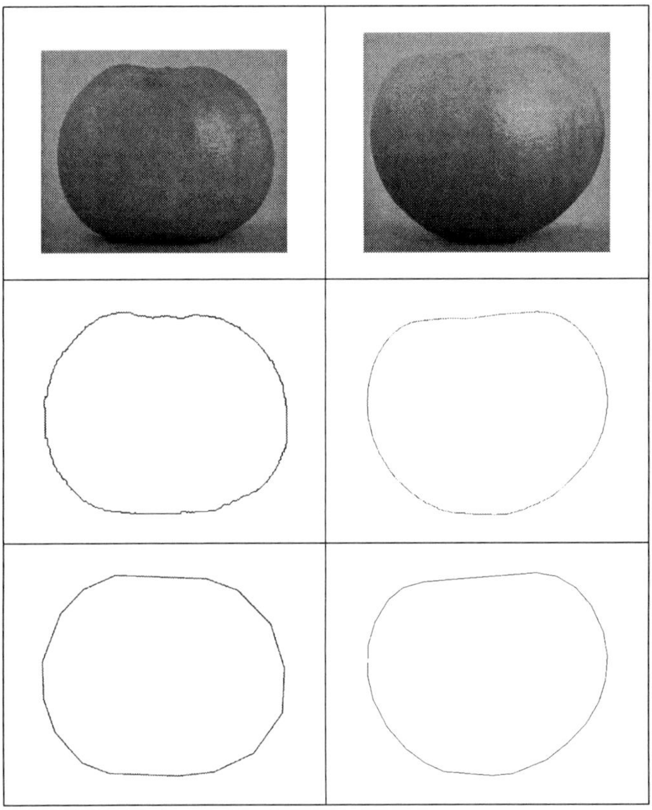

Table A.14: Rother Fenchelapfel

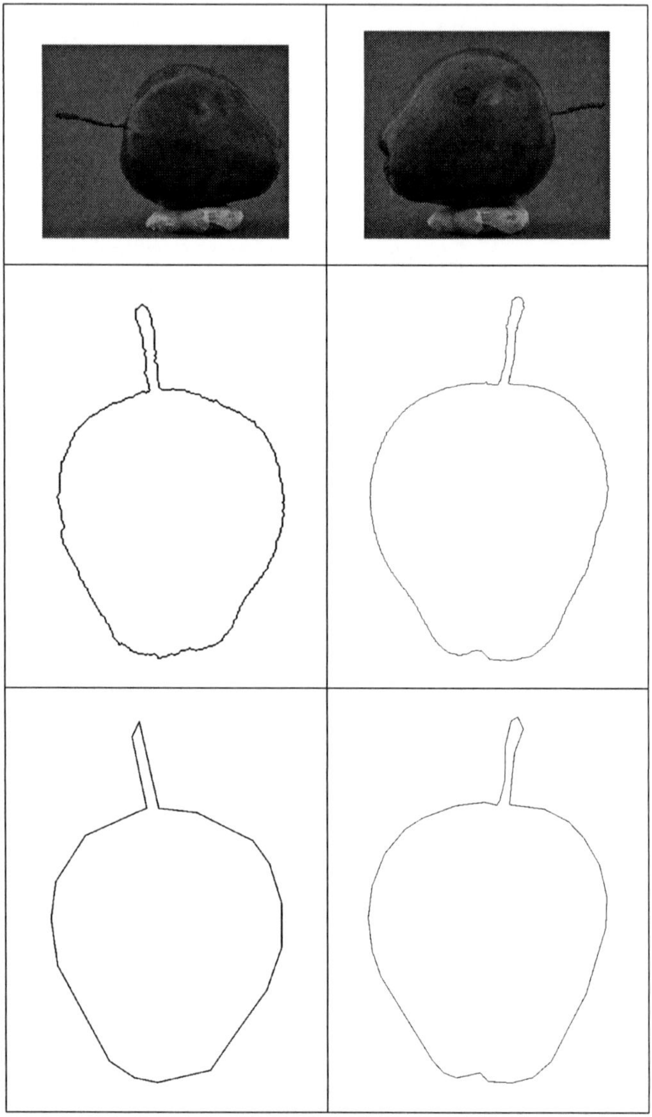

Table A.15: Rother Herbstcalville

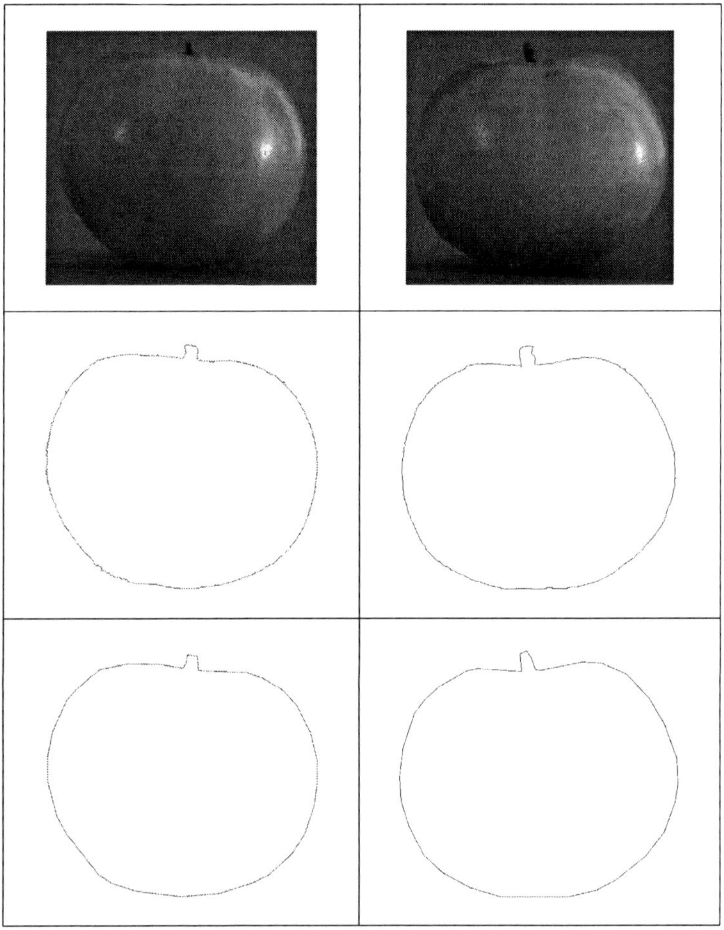

Table A.16: Rother Stettiner

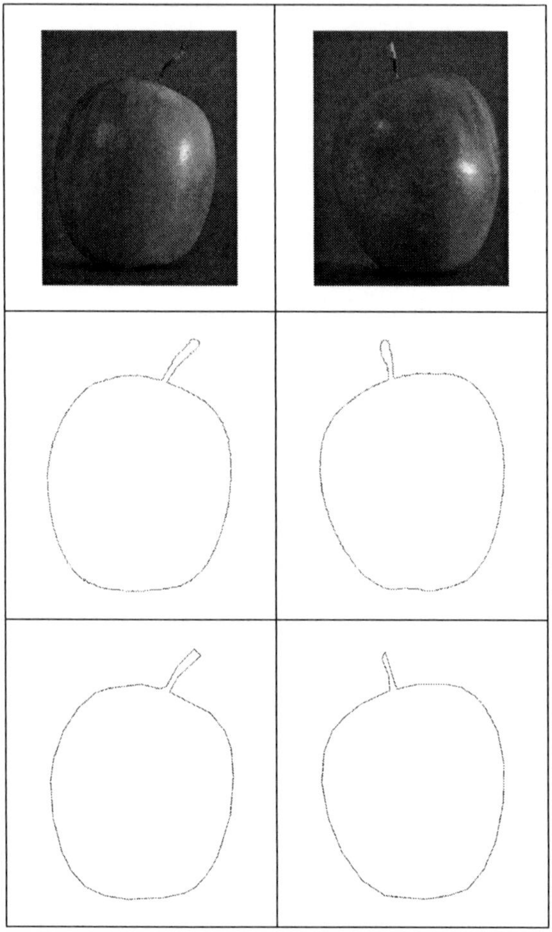

Table A.17: Rother Taubenapfel

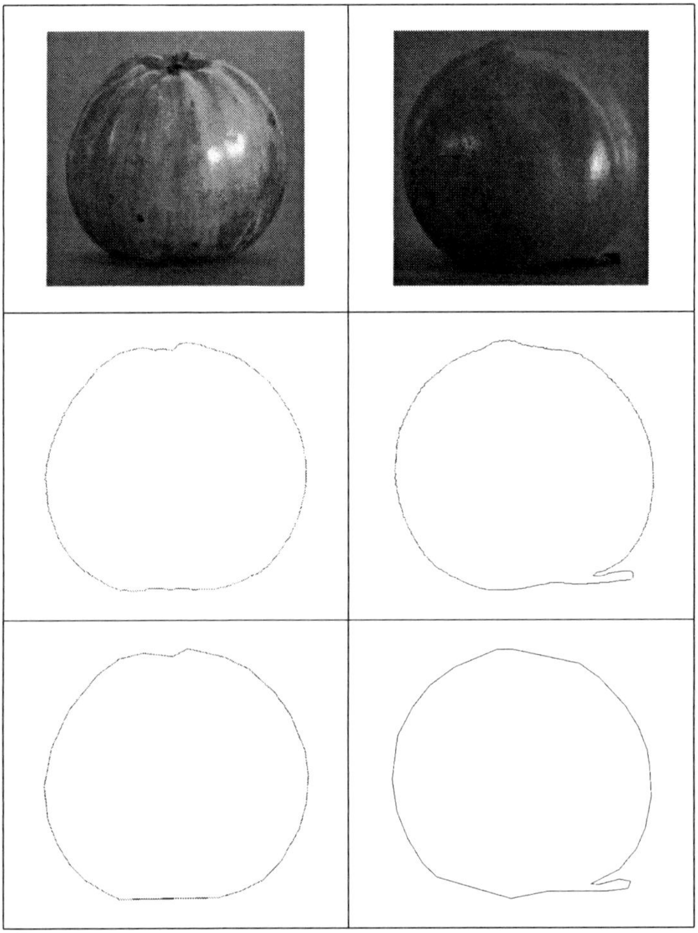

Table A.18: Rother Wintercalville

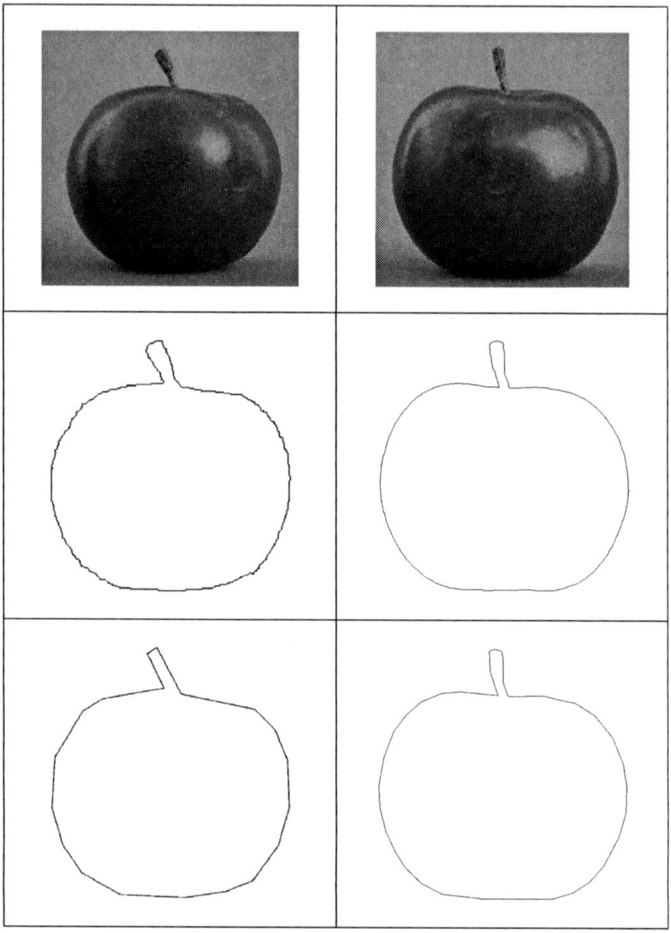

Table A.19: Schwarzer Borsdorferapfel

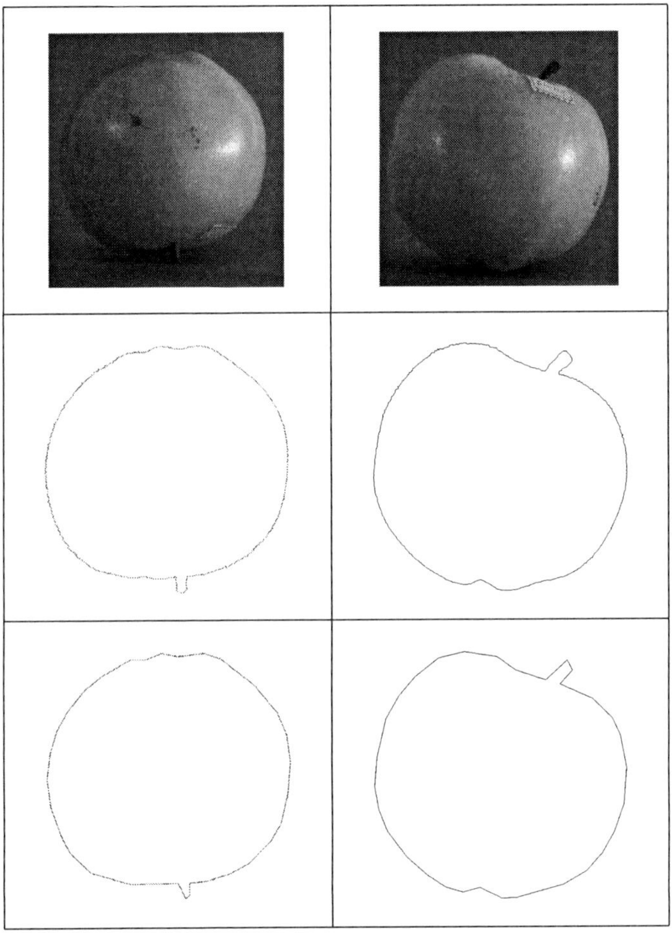

Table A.20: Sommer Zuckerapfel

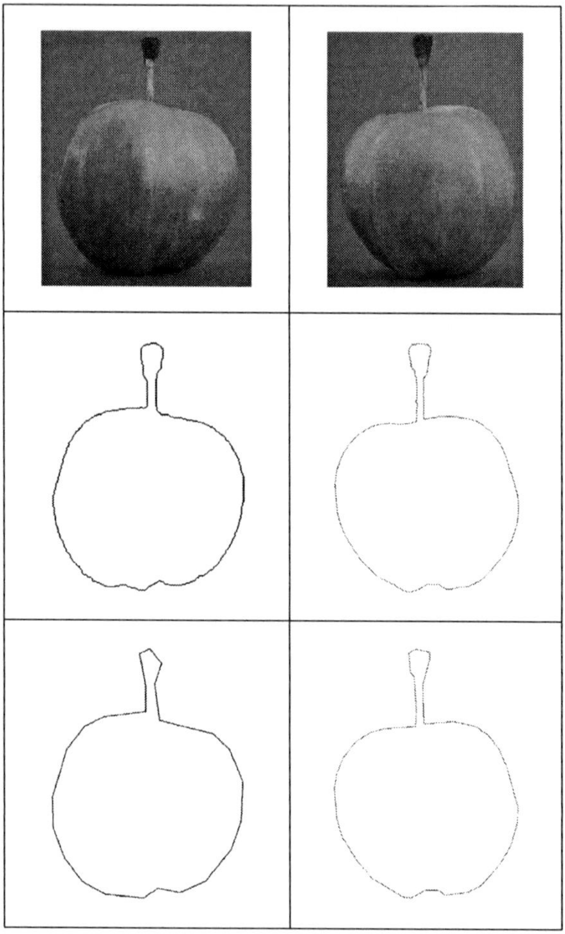

Table A.21: Veilchenapfel

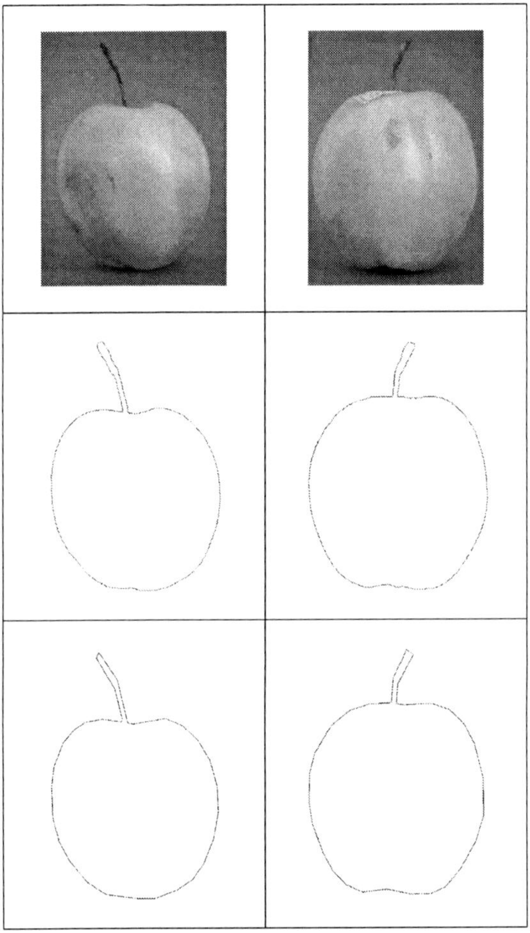

Table A.22: Weiße Herbstkalville

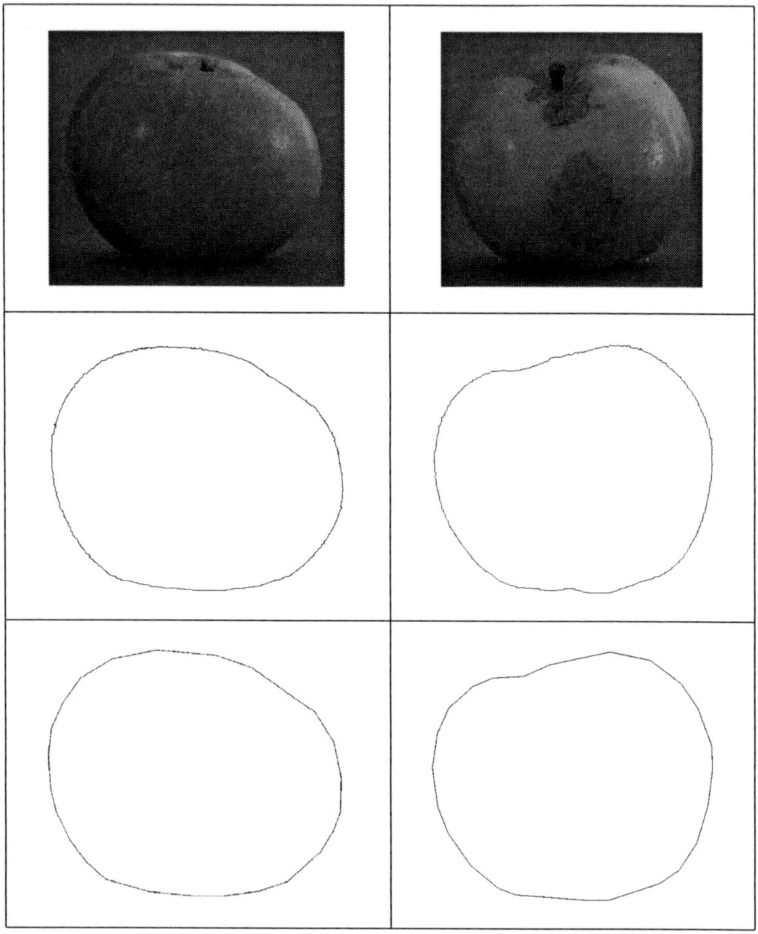

Table A.23: Weißer Maatapfel

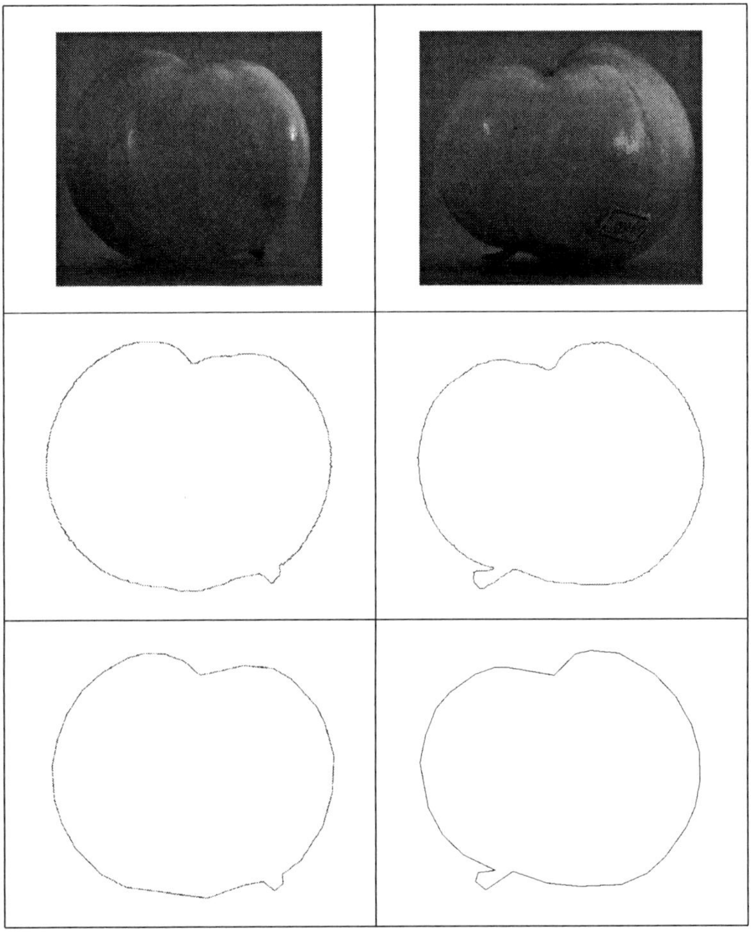

Table A.24: Weisser Sommercalvil

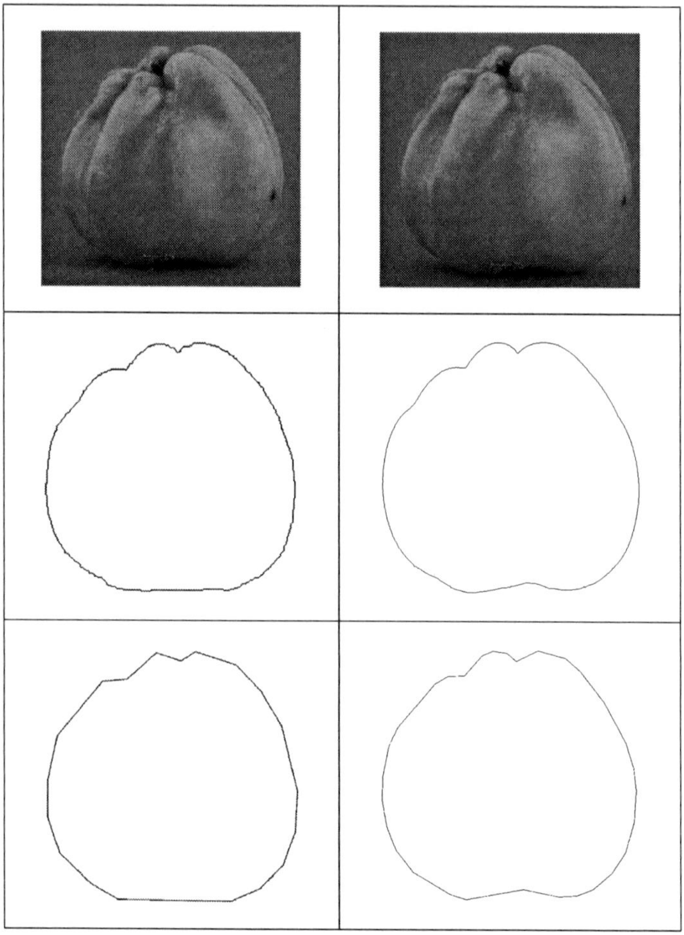

Table A.25: Weisser Winterkalvill

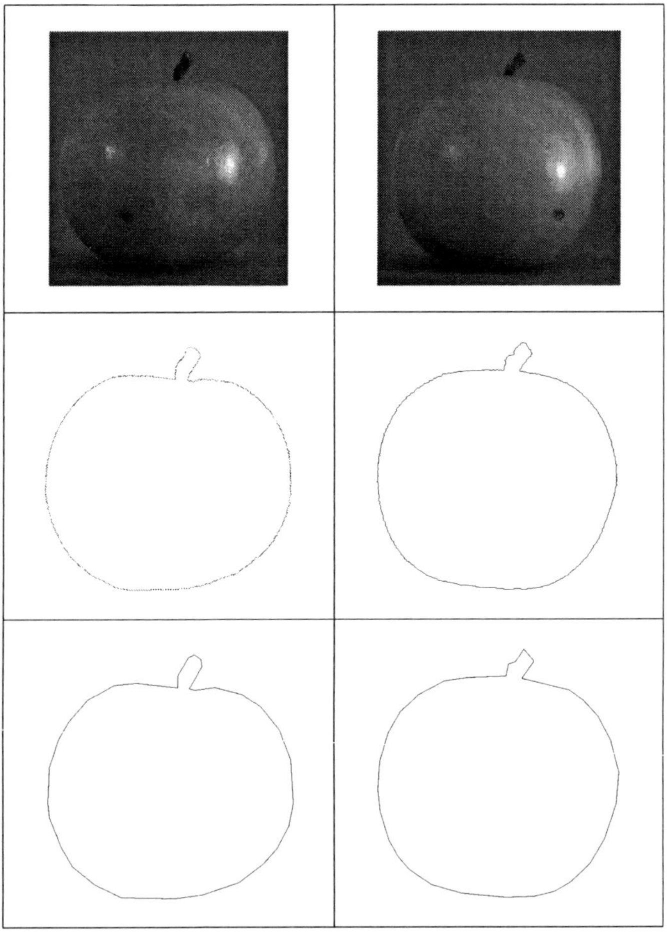

Table A.26: Zwei Jahre dauernde Renette

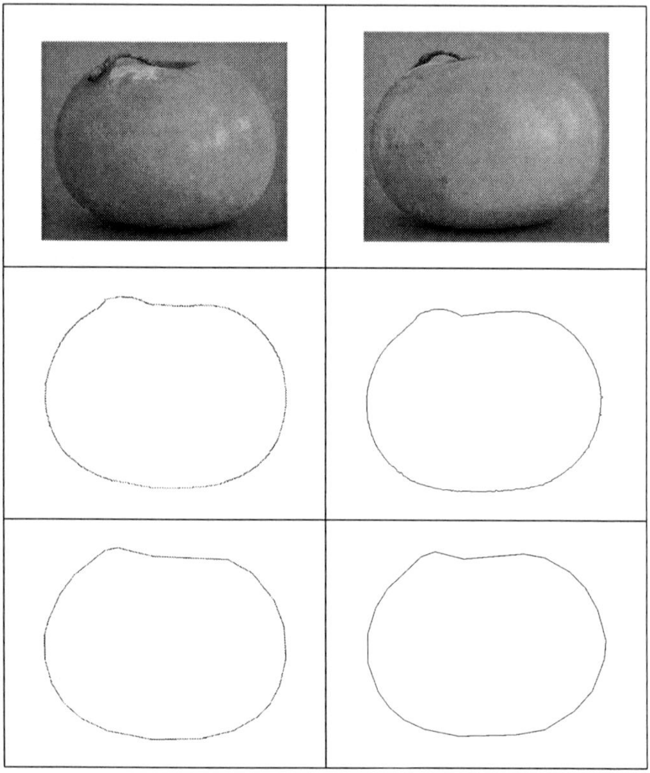

Table A.27: Zwiebelborsdorferapfel

Appendix B

Queries and reference sets

The six queries of the evaluation scenario described in chapter 5 are given on the following pages together with the corresponding reference sets; the numbers of the fruits are those given in Table 5.1, while their order corresponds to the order in the result set of the qualitative approach. On the left in each case is the original sketch, while the right-hand part of each figure shows the polygonal approximation. As in the Bertuch-collection (Appendix A), the maximal error between the original contour and the polygonal shape is 5 pixels. The qualitative matrix of each polygon is given on the following page.

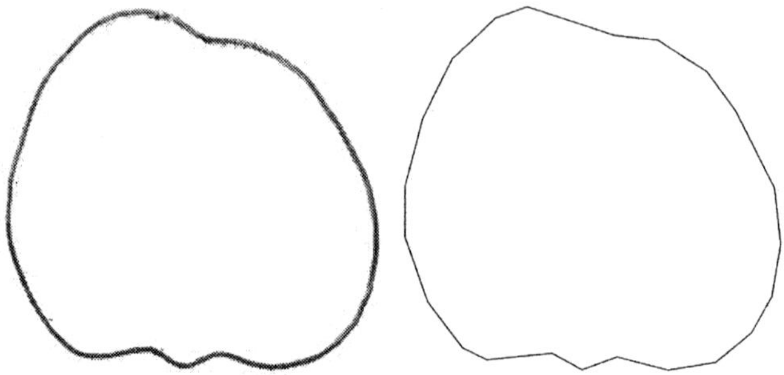

Figure B.1: The first query

Table B.1: The first sketch and its precision and recall

Fruit No.	Precision optimal	Precision non-optimal	Precision quantitative	Recall
7 I	1.00	1.00	0.10	0.14
1 I	1.00	1.00	0.15	0.28
20 I	0.60	0.60	0.21	0.42
21 I	0.66	0.28	0.18	0.57
21 II	0.71	0.33	0.18	0.71
25 I	0.75	0.33	0.17	0.85
18 I	0.35	0.30	0.13	1.00

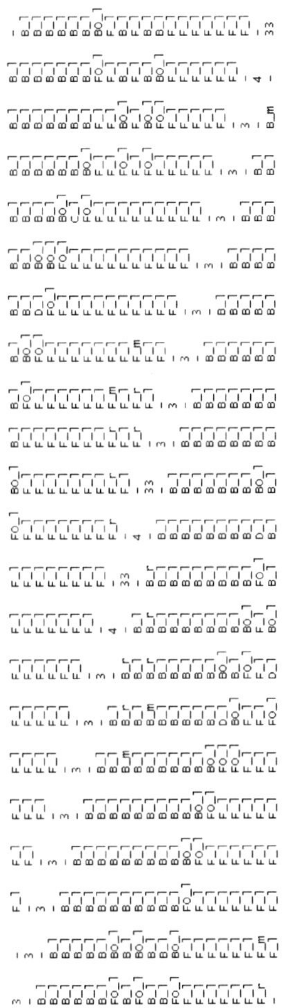

Figure B.2: Qualitative matrix of query one

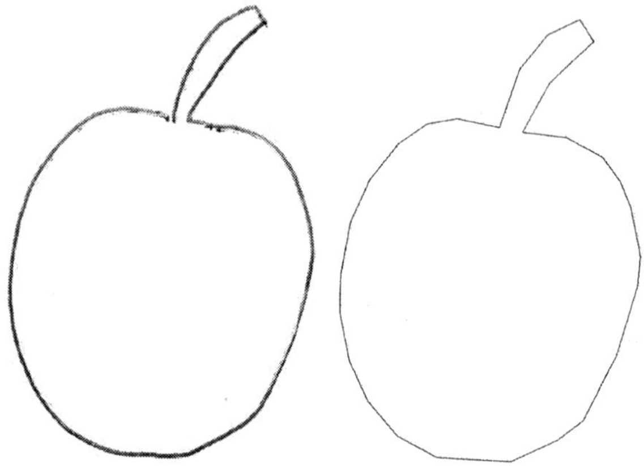

Figure B.3: The second query

Table B.2: The second sketch and its precision and recall

Fruit No.	Precision optimal	Precision non-optimal	Precision quantitative	Recall
8 II	1.00	1.00	0.16	0.10
12 I	1.00	1.00	0.15	0.20
15 I	1.00	1.00	0.20	0.30
12 II	1.00	0.57	0.21	0.40
8 I	0.62	0.50	0.23	0.50
15 II	0.66	0.50	0.26	0.60
17 I	0.70	0.53	0.24	0.70
17 II	0.72	0.57	0.22	0.80
22 I	0.75	0.50	0.23	0.90
22 II	0.76	0.52	0.24	1.00

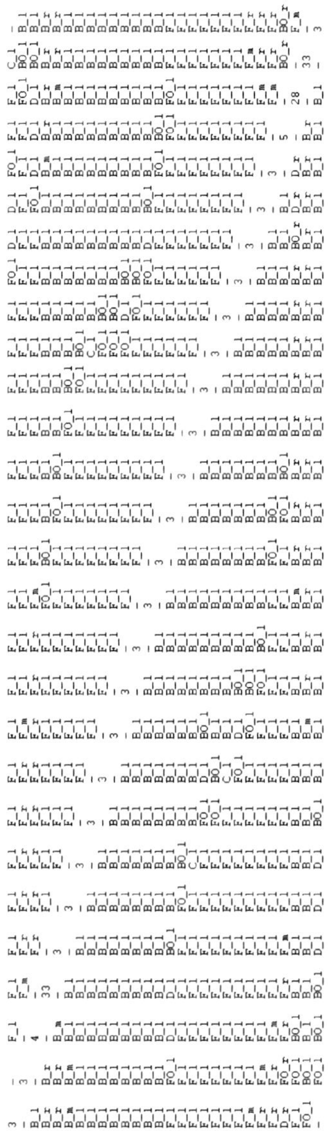

Figure B.4: Qualitative matrix of query two

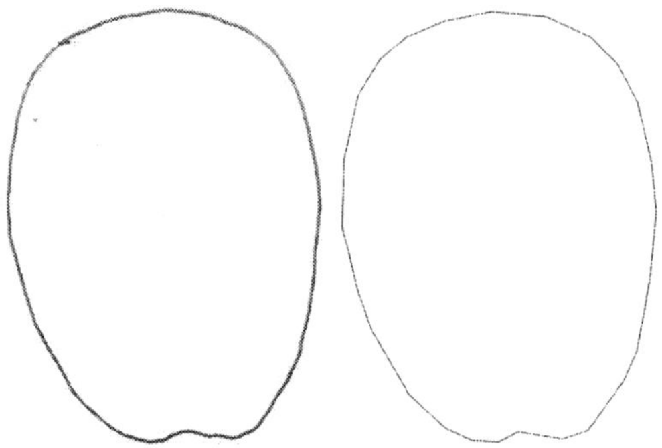

Figure B.5: The third query

Table B.3: The third sketch and its precision and recall

Fruit No.	Precision optimal	Precision non-optimal	Precision quantitative	Recall
3 I	0.25	0.25	0.50	0.07
7 II	0.40	0.40	0.50	0.14
8 I	0.50	0.50	0.50	0.21
15 II	0.57	0.50	0.19	0.28
17 I	0.62	0.55	0.22	0.35
17 II	0.66	0.60	0.25	0.42
22 I	0.70	0.50	0.24	0.50
22 II	0.72	0.53	0.26	0.57
5 I	0.50	0.50	0.23	0.64
5 II	0.52	0.52	0.23	0.71
12 I	0.45	0.40	0.25	0.78
12 II	0.48	0.42	0.24	0.85
15 I	0.50	0.44	0.25	0.92
7 I	0.46	0.46	0.25	1.00

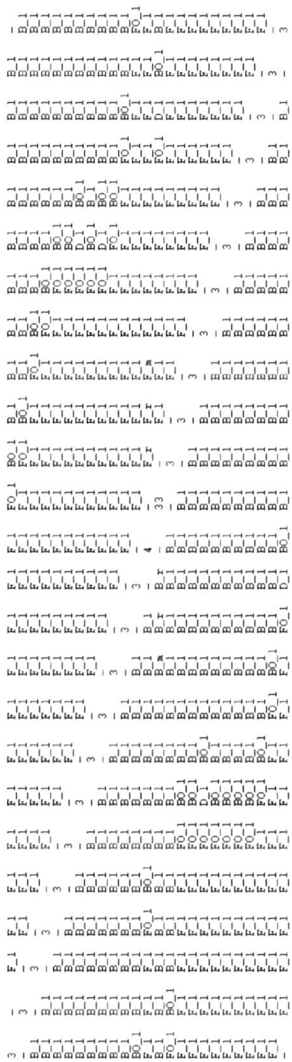

Figure B.6: Qualitative matrix of query three

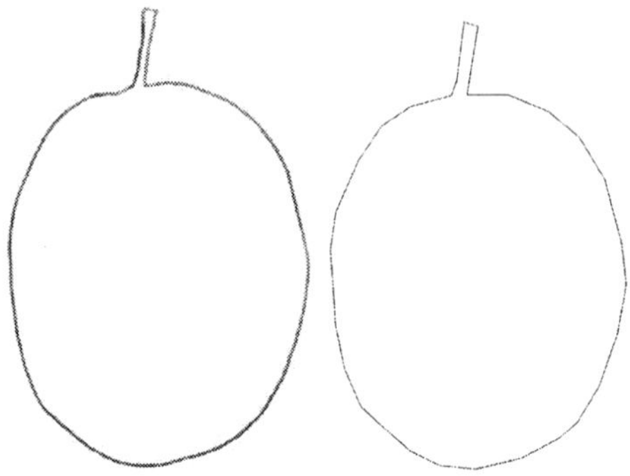

Figure B.7: The fourth query

Table B.4: The fourth sketch and its precision and recall

Fruit No.	Precision optimal	Precision non-optimal	Precision quantitative	Recall
8 II	1.00	1.00	0.25	0.09
12 I	1.00	1.00	0.10	0.18
15 I	1.00	1.00	0.14	0.27
11 I	1.00	0.80	0.13	0.36
11 II	1.00	0.83	0.15	0.45
12 II	1.00	0.85	0.17	0.54
15 II	0.87	0.58	0.17	0.63
17 I	0.88	0.61	0.17	0.72
17 II	0.90	0.64	0.19	0.81
22 I	0.90	0.55	0.20	0.90
22 II	0.91	0.57	0.22	1.00

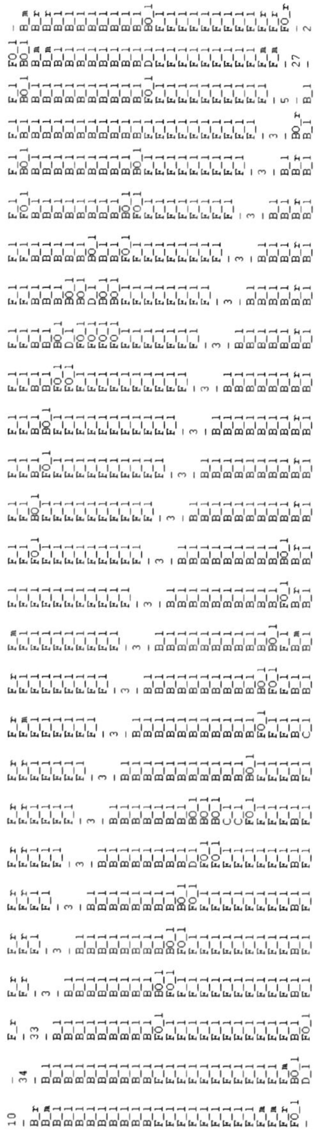

Figure B.8: Qualitative matrix of query four

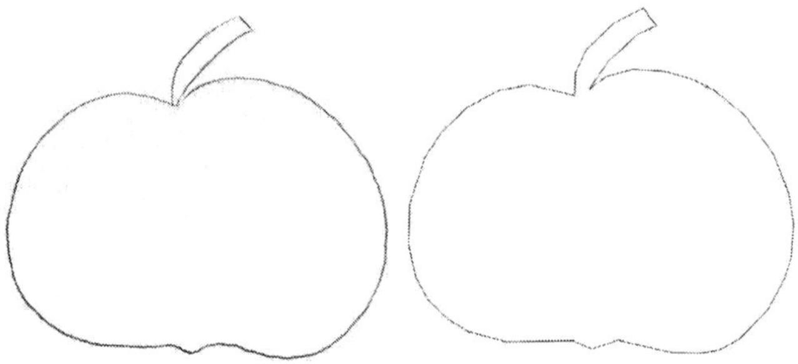

Figure B.9: The fifth query

Table B.5: The fifth sketch and its precision and recall

Fruit No.	Precision optimal	Precision non-optimal	Precision quantitative	Recall
13 I	1.00	0.20	1.00	0.16
19 I	1.00	0.33	0.40	0.33
20 I	0.42	0.42	0.42	0.50
14 II	0.50	0.44	0.12	0.66
19 II	0.41	0.31	0.10	0.83
24 I	0.46	0.35	0.11	1.00

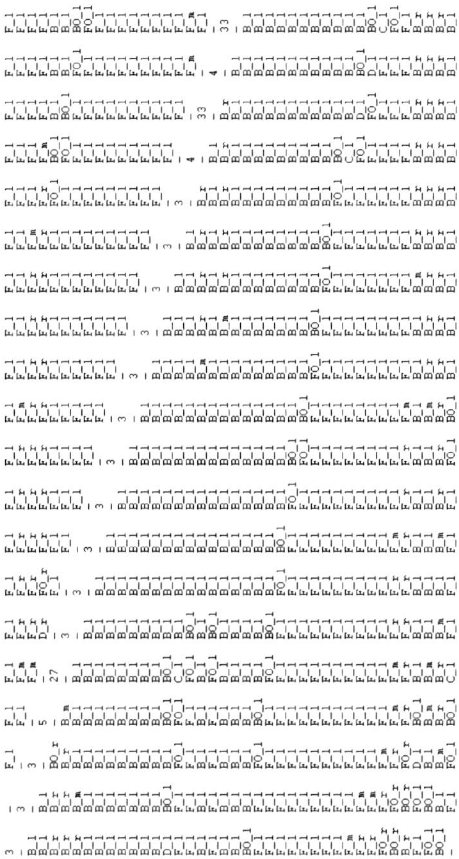

Figure B.10: Qualitative matrix of query five

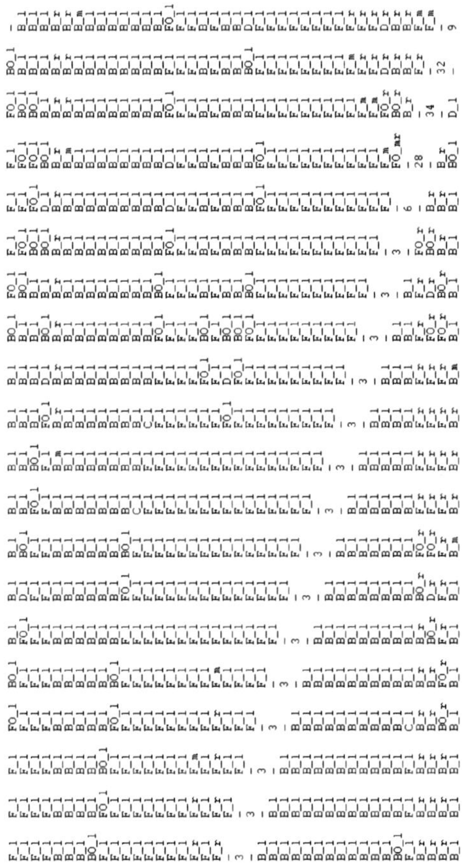

Figure B.11: Qualitative matrix of query five - continued

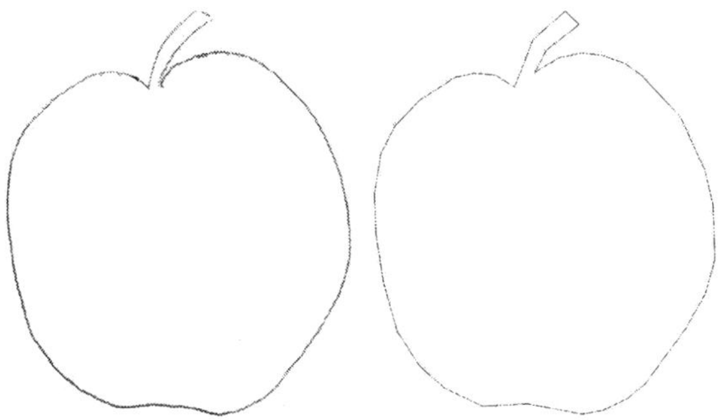

Figure B.12: The sixth query

Table B.6: The sixth sketch and its precision and recall

Fruit No.	Precision optimal	Precision non-optimal	Precision quantitative	Recall
3 I	1.00	1.00	0.20	0.12
7 II	1.00	1.00	0.25	0.25
8 I	1.00	1.00	0.33	0.37
15 II	1.00	0.80	0.17	0.50
22 I	1.00	0.45	0.16	0.62
22 II	1.00	0.50	0.14	0.75
8 II	0.41	0.41	0.14	0.87
19 II	0.40	0.40	0.15	1.00

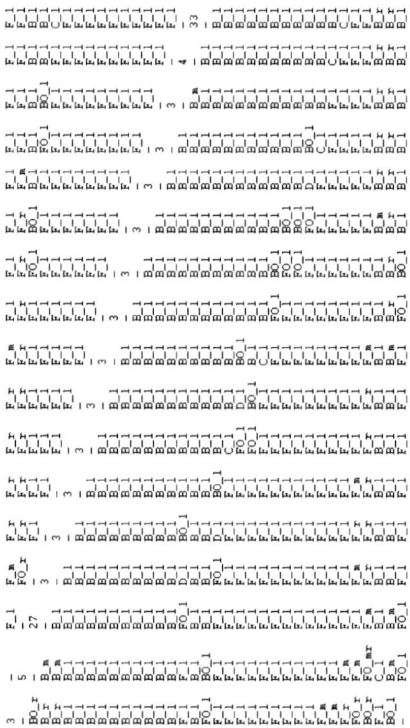

Figure B.13: Qualitative matrix of query six

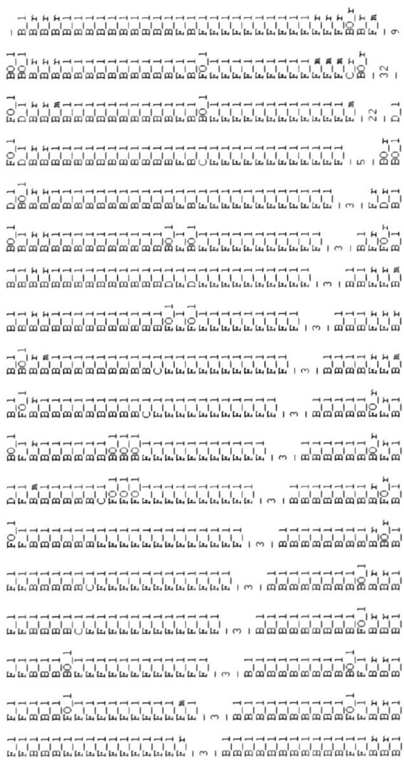

Figure B.14: Qualitative matrix of query six - continued

Bibliography

Allen, J. F. (1983). Maintaining knowledge about temporal intervals. *Communications of the ACM, 26*(11), 832-843.

Alvarado, C. & Davis, R. (2001). Resolving Ambiguities to Create a Natural Computer-Based Sketching Environment. In B. Nebel (Ed.), *Proceedings of the Seventeenth International Joint Conference on Artificial Intelligence, IJCAI 2001* (p. 1365-1374). Morgan Kaufmann.

Arnheim, R. (1978). *Kunst und Sehen. Eine Psychologie des schöpferischen Auges.* Walter de Gruyter.

Attneave, F. (1954). Some informational aspects of visual perception. *Psychological Review, 61*, 183-193.

Baeza-Yates, R. A. & Ribeiro-Neto, B. A. (1999). *Modern Information Retrieval.* ACM Press / Addison-Wesley.

Bartlett, F. C. (1932). *Remembering.* Cambridge University Press.

Byrne, R. W. (1979). Memory for urban geography. *Quarterly Journal of Experimental Psychology, 31*, 147-154.

Clementini, E. & Felice, P. D. (1997). A Global Framework for Qualitative Shape Description. *GeoInformatica, 1*(1), 11-27.

Cohen, R., Baldwin, L. M. & Sherman, R. C. (1978). Cognitive maps of a naturalistic setting. *Child Development, 49*, 1216-1218.

Cohn, A. G. (1995). A hierarchical representation of qualitative shape based on connection and convexity. In A. M. Frank & W. Kuhn (Eds.), *COSIT 1995, Spatial Information Theory* (p. 311-326). Springer-Verlag.

Cohn, A. G. & Hazarika, S. M. (2001). Qualitative Spatial Representation and Reasoning: An Overview. *Fundamenta Informaticae, 43*, 2-32.

Cordella, L. P. & Vento, M. (2000). Symbol recognition in documents: a collection of techniques? *International Journal on Document Analysis and Recognition, 3*, 73-88.

Couclelis, H., Golledge, R. G., Gale, N. & Tobler, W. (1987). Exploring the anchor-point hypothesis of spatial cognition. *Journal of Environmental Psychology, 7*, 99-122.

DiCarlo, J. J. & Maunsell, H. R. (2000). Form representation in monkey inferotemporal cortex is virtually unaltered by free viewing. *Nature neuroscience, 3(8)*, 814-821.

Douglas, D. H. & Peucker, T. K. (1973). Algorithms for the reduction of the number of points required to represent a digitized line or its caricature. *The Canadian Cartographer*, *10(2)*, 112-122.

Egenhofer, M. (1997). Query Processsing in Spatial-Query-by-Sketch. *Journal of Visual Languages and Computing*, *8*, 403-424.

Egenhofer, M. & Franzosa, R. (1991). Point-set topological spatial relations. *International Journal of Geographical Information Systems*, *5*(2), 161-174.

Egenhofer, M. & Herring, J. (1991). Categorizing binary topological relationships between regions, lines, and points in Geographic Databases. *Technical Report, Department of Surveying Engineering, University of Maine, Orono.*

Ferguson, R. W. & Forbus, K. D. (2002). A Cognitive Approach to Sketch Understanding. In *AAAI Spring Symposium on Sketch Understanding.* American Association for Artificial Intelligence.

Freksa, C. (1992a). Temporal reasoning based on semi-intervals. *Artificial Intelligence*, *94*, 199-227.

Freksa, C. (1992b). Using Orientation Information for Qualitative Spatial Reasoning. In A. M. Frank, I. Campari & U. Formentini (Eds.), *Theories and Methods of Spatio-Temporal Reasoning in Geographic Space.* Springer-Verlag.

Freksa, C. & Zimmermann, K. (1992). On the Utilization of Spatial Structures for Cognitively Plausible and Efficient Reasoning. In *IEEE International Conference on Systems, Man and Cybernetics* (p. 261-266). Chicago.

Freundschuh, S. & Egenhofer, M. (1997). Human Conceptions of Spaces: Implications for GIS. *Transactions in GIS*, *4*, 361-375.

Galton, A. & Meathrel, R. C. (1999). Qualitative Outline Theory. In T. Dean (Ed.), *Proc. of 16th IJCAI-99* (p. 1061-1066). Stockholm, Sweden: Morgan Kaufmann, Inc.

Goldenberg, G. (1998). *Neuropsychologie.* Stuttgart: Gustav Fischer.

Gottfried, B. (2002). Tripartite Line Tracks. In K. Wojciechowski (Ed.), *International Conference on Computer Vision and Graphics* (p. 288-293). Zakopane, Poland.

Gottfried, B. (2003a). Tripartite Line Tracks, Bipartite Line Tracks. In A. Günter, R. Kruse & B. Neumann (Eds.), *KI 2003: Advances in Artificial Intelligence* (p. 535-549). Hamburg, Germany: Springer-Verlag.

Gottfried, B. (2003b). Tripartite Line Tracks, Qualitative Curvature Information. In W. Kuhn, M. Worboys & S. Timpf (Eds.), *Spatial Information Theory: Foundations of Geographic Information Science, Int. Conference COSIT 2003* (p. 101-117). Ittingen, Switzerland: Springer-Verlag.

Gottfried, B. (2004a). Reasoning about Intervals in Two Dimensions. In W. Thissen, P. Wieringa, M. Pantic & M. Ludema (Eds.), *IEEE Int. Conference on Systems, Man and Cybernetics* (p. 5324-5332). The Hague, The Netherlands: Omnipress.

Gottfried, B. (2004b). Singularities in Qualitative Reasoning. In B. Buchberger & J. Campbell (Eds.), *AISC 2004, 7 th Int. Conference on Artificial Intelligence and Symbolic Computation* (p. 276-280). Castle of Hagenberg, Linz, Austria: Springer-Verlag.

Gregory, R. L. (1997). Visual illusions classified. *Trends in Cognitive Science, 1 (5)*, 190-194.

Gross, M. D. (1998). The proverbial back of an envelope. *IEEE Intelligent Systems, 13*, 10-13.

Guesgen, H. (1989). *Spatial reasoning based on Allen's temporal logic.* International Computer Science Institute, Berkley. (TR-89-049)

Hammond, T. & Davis, R. (2003). LADDER: A Language to Describe Drawing, Display, and Editing in Sketch Recognition. In *Proceedings of the Ninth International Joint Conference on Artificial Intelligence, IJCAI 2003.*

Hernandez, D. (1992). Diagrammatical Aspects of Qualitative Representations of Space. In *AAAI Spring Symposium on Reasoning with Diagrammatic Representations.* (p. 225-228). Stanford.

Hernandez, D., Clementini, E. & Di Felice, P. (1995). Qualitative Distances. In A. U. Frank & W. Kuhn (Eds.), *Spatial Information Theory: Foundations of Geographic Information Science, Int. Conference COSIT 1995* (p. 45-57). Springer.

Hirtle, S. C. & Jonides, J. (1985). Evidence of hierarchies in cognitive maps. *Memory and Cognition, 13*, 208-217.

Holyoak, K. J. & Mah, W. A. (1982). Cognitive reference points in judgments of symbolic magnitude. *Cognitive Psychology, 14*, 328-352.

Horng, J. & Li, J. T. (2002). An automatic and efficient dynamic programming algorithm for polygonal approximation of digital curves. *Pattern Recognition Letters, 23*, 171-182.

Hse, H., Shilman, M. & Newton, A. R. (2004). Robust sketched symbol fragmentation using templates. In J. Vanderdonckt, N. J. Nunes & C. Rich (Eds.), *Proceedings of the 2004 International Conference on Intelligent User Interfaces, January 13-16, 2004* (p. 156-160). Funchal, Madeira, Portugal: ACM.

Itten, J. (1963). *Gestaltungs- und Formenlehre — Mein Vorkurs am Bauhaus und später.* Ravensburger Buchverlag.

Jungert, E. (1993). Symbolic spatial reasoning on object shapes for qualitative matching. In A. U. Frank & L. Campari (Eds.), *COSIT 1993, Spatial Information Theory: A Theoretical Basis for GIS.* (p. 444-462). Springer-Verlag.

Kaiser, G. R. & Seifert, S. (2000). *Friedrich Justin Bertuch (1747-1822); Verleger, Schriftsteller und Unternehmer im klassischen Weimar.* Max Niemeyer Verlag, 2000.

Kennedy, J. M. (2003). Drawings from Gaia, a blind girl. *Perception, 32*, 321-340.

Ladkin, P. & Maddux, R. (1994). On binary constraint problems. *Journal of the Association for Computing Machinery, 41(3)*, 435-469.

Landay, J. A. (1998). Informal User Interface For Natural Human-Computer Interaction. *IEEE Intelligent Systems, 13*, 14-16.

Landay, J. A. & Myers, B. A. (2001). Sketching Interfaces: Toward More Human Interface Design. *IEEE Computer, 34*, 56-64.

Landerer, C. (2000). *Kunstgeschichte als Kognitionsgeschichte: Ein Beitrag zur genetischen Kulturpsychologie.* Doctoral Dissertation, University of Salzburg, Austria.

Long Jr., A. C., Landay, J. A. & Rowe, L. A. (1997). *PDA and gesture use in practice: Insights for designers of pen-based user interfaces.*

Mandelbrot, B. B. (1986). Fractals and the Rebirth of Iteration Theory. In H.-O. Peitgen & P. H. Richter (Eds.), *The Beauty of Fractals: Images of Complex Dynamical Systems* (p. 151-160). Springer.

Marr, D. (1982). *Vision - A Computational Investigation into the Human Representation and Processing of Visual Information.* New York: W. H. Freeman and Company.

Matsakis, P. & Wendling, L. (1999). A new way to represent the relative position between areal objects. *IEEE Transactions on Pattern Analysis and Machine Intelligence, 21*, 634-643.

Mäuser, M. (2002). Venusbrust und Seidenhemdchen. *aviso - Zeitschrift für Wissenschaft und Kunst in Bayern, 4*, 42-45.

Meathrel, R. C. & Galton, A. (2000). Qualitative Representation of Planar Outlines. In W. Horn (Ed.), *Proceedings of 14th ECAI* (p. 224-228). IOS Press.

Meathrel, R. C. & Galton, A. (2001). A Hierarchy of Boundary-based Shape Descriptors. In B. Nebel (Ed.), *Proceedings of the Seventeenth International Joint Conference on Artificial Intelligence, IJCAI 2001* (p. 1359-1364). Seattle, Washington, USA: Morgan Kaufmann.

Milbrath, C. (1998). *Patterns of Artistic Development in Children: Comparative Studies of Talent.* Cambridge: Cambridge University Press.

Milgram, S. & Jodelet, D. (1976). Psychological maps of Paris. In *Environmental psychology* (p. 104-124). New York: Proshansky, H. M. and Ittelson, W. H. and Rivlin, L. G.

Mitzias, D. A. & Mertzios, B. G. (1994). Shape Recognition With A Neural Classifier Based On A Fast Polygonal Approximation Technique. *Pattern Recognition, 27*, 627-637.

Moratz, R., Renz, J. & Wolter, D. (2000). Qualitative Spatial Reasoning about Line Segments. In *ECAI 2000 Proceedings of the 14th European Conference on Artificial Intelligence.* Amsterdam: IOS Press.

Norman, J. F., Phillips, F. & Ross, H. E. (2001). Information concentration along the boundary contours of naturally shaped solid objects. *Perception, 30*, 1285-1294.

Okabe, A. & Masuyama, A. (2001). An exploratory method for qualitative trend curve analysis — a robust method against low quality data. *International Journal of Geographical Information Science, 15*, 65-76.

Palmer, S. (1999). *Vision Science, From Photons to Phenomenology.* MIT Press.

Petzold, H. (1982). *Apfelsorten.* Neumann Verlag.

Randell, D. A., Cui, Z. & Cohn, A. G. (1992). A spatial logic based on regions and connection. In *Proc 3rd Int. Conf. on Knowledge Representation and Reasoning* (p. 165-176). San Mateo: Morgan Kaufman.

Renz, J. (2001). A Spatial Odyssey of the Interval Algebra: 1. Directed Intervals. In B. Nebel (Ed.), *Proceedings of the Seventeenth International Joint Conference on Artificial Intelligence, IJCAI 2001, Seattle, Washington, USA, August 4-10, 2001* (p. 51-56). Morgan Kaufmann.

Rosin, P. L. (2003). Assessing the behaviour of polygonal approximation algorithms. *Pattern Recognition, 36*, 505-518.

Sadalla, E. K., Burroughs, W. J. & Staplin, L. I. (1980). Reference points in spatial cognition. *Journal of Experimental Psychology: Human Learning and Memory, 5*, 516-528.

Sadalla, E. K. & Magel, S. G. (1980). The perception of traversed distance. *Environment and Behaviour, 12*, 65-79.

Schlieder, C. (1993). Representing visible locations for qualitative navigation. In N. P. Carrete & M. G. Singh (Eds.), *Qualitative reasoning and decision technologies* (p. 523-532).

Schlieder, C. (1995a). Ordering information and symbolic projection. In S. K. Chang & al. (Eds.), *Intelligent image database systems.* Singapore: World Scientific.

Schlieder, C. (1995b). Reasoning about ordering. In A. Frank & W. Kuhn (Eds.), *International conference on spatial information theory* (p. 341-349). Springer.

Schlieder, C. (1996). Qualitative Shape Representation. In P. Burrough & A. M. Frank (Eds.), *Geographic objects with indeterminate boundaries* (p. 123-140). London: Taylor & Francis.

Schlieder, C. & Hagen, C. (2000). Interactive Layout Generation with a Diagrammatic Constraint Language. In C. e. a. Freksa (Ed.), *Spatial Cognition II* (p. 198-211). Springer, LNAI 1849.

Serra, J. (1983). *Image analysis and mathematical morphology.* Academic Press.

Sezgin, T., Stahovich, T. & Davis, R. (2001). Sketch Based Interfaces: Early Processing for Sketch Understanding. In *Proceedings of 2001 Perceptive User Interfaces Workshop (PUI'01)* (p. Sezgin). Orlando FL, USA.

Skubic, M., Blisard, S., Bailey, J. A., C. Adams & Matsakis, P. (2004). Qualitative Analysis of Sketched Route Maps: Translating a Sketch Into Linguistic Descriptions. *IEEE Transactions on Systems, Man, and Cybernetics - Part B: Cybernetics, 34*, 1275-1282.

Stadler, M. A., Richter, P. H. & Pfaff, P., S. Kruse. (1991). Attractors and perceptual field dynamics of homogeneous stimulus areas. *Psychological Research, 53*, 102-112.

Stadler, M. A. & Wildgen, W. (1987). Ordnungsbildung beim Verstehen und bei der Reproduktion von Texten. *Siegener Periodicum zur Internationalen empirischen Literaturwissenschaft, 6*, 101-144.

Stahovich, T. F. (1998). The Engineering Sketch. *IEEE Intelligent Systems, 13*, 17-19.

Stevens, A. & Coupe, P. (1978). Distortions in judged spatial relations. *Cognitive Psychology, 13*, 422-437.

Thorndyke, P. W. (1981). Distance estimation from cognitive maps. *Cognitive Psychology, 13*, 526-550.

Tobler, W. (1970). A computer movie simulating urban growth in the Detroit region. *Economic Geography, 46 (4)*, 360-371.

Tversky, B. (1981). Distortions in memory for maps. *Cognitive Psychology, 13*, 407-433.

Tversky, B. (1992). Distortions in cognitive maps. *Geoforum, 23*, 131-138.

Tversky, B., Zacks, J., Lee, P. & Heiser, J. (2000). Lines, Blobs, Crosses and Arrows: Diagrammatic Communication with Schematic Figures. In M. Anderson, P. Cheng & V. Haarslev (Eds.), *Diagrams 2000* (Vol. 1889, p. 221-230). Springer-Verlag.

Willats, J. (1997). *Art and Representation.* Princeton: Princeton University Press.

Zimmermann, K. & Freksa, C. (1996). Qualitative Spatial Reasoning Using Orientation, Distance, and Path Knowledge. *Applied Intelligence, 6*, 49–58.